24 0392690 6

AF598280

Imperial College &
Science Museum Libraries

London SW7 2AZ Issue Desk Tel. 0171-594 8810

THREE WEEK LOAN
To be returned by the last date stamped.

A fine may be charged if the book is returned late.

30 SEP 1997
10 NOV 1997
25 NOV 1997
05 DEC 1997
12 JAN 1998
24 FEB 1998
29 APR 1998
1.6.98
11 JAN 1999
-5 MAR 1999
28 APR 1999
7 MAY 1999
-9 JUN 1999
30 SEP 1999
11 NOV 1999
03 MAY 2000
-1 MAR 2001
02 APR 2001
18 MAY 2001
29 JUN 2001
-9 JAN 2002
23 DEC 2003
-5 JAN 2004
-1 NOV 2004
30 SEP 2010

BASIC CORROSION TECHNOLOGY FOR SCIENTISTS AND ENGINEERS

Basic Corrosion Technology

for Scientists and Engineers

Second Edition

Einar Mattsson

THE INSTITUTE OF MATERIALS

7/97 620.193 MAT

Book 636
This edition first published in 1996 by
The Institute of Materials
1 Carlton House Terrace
London SW1Y 5DB

First English language edition published in 1989 by
Ellis Horwood Ltd

This edition © The Institute of Materials 1996
Illustrations © Einar Mattsson

ISBN 1 86125 011 8

All Rights Reserved. No part of this publication may be reproduced, stored in a retrieval system, or transmitted, in any form or by any means, electronic, mechanical, photocopying, recording or otherwise, without the permission of The Institute of Materials, 1 Carlton House Terrace, London SW1Y 5DB, UK.

Neither the publishers, nor the Editors and the authors and their employing organisations, not any person acting on behalf of any of them (a) makes any warranty, express or implied, with respect to the use of any information, apparatus, method or process disclosed in this book or that such use may not infringe privately owned right or (b) assumes any liabilities with respect to the use of, or for damages resulting from the use of, any information, apparatus, method or process disclosed in this book, and the use of any materials or methods is solely at the risk of the user.

Typeset, printed and bound in the UK by
Bourne Press, Bournemouth

Table of contents

Preface

The first edition of this book was published in 1989 by Ellis Horwood Ltd, Chichester, UK. When this edition had been sold out in 1994, there was still an interest in the book. For that reason The Institute of Materials, London has decided to publish a second edition, updated with respect to recent developments.

When preparing the manuscript for the second edition, I have received assistance from Kaija Eistrat, Louise Hult, Bror Sederholm and Bengt Sundvall, Swedish Corrosion Institute, from Christofer Leygraf, Royal Institute of Technology, Stockholm, and from Christer Ornby, MMS, Stockholm. Many thanks for this valuable support.

I hope the book will be of continued interest and value to the readers.

Stocksund
15 December 1995

Einar Mattsson

Preface to First Edition

The aim of this book on basic corrosion technology for scientists and engineers is to give a survey of the corrosion of metals, its damage to the community and possible countermeasures. Efforts have been made to write a book easy to read, even for persons without deep knowledge of chemisty.

The book is based on my experience from 15 years as corrosion metallurgist in Swedish industry and from 17 years as director of research or leader of the Swedish Corrosion Institute.

When writing the book I have had the privilege of discussion with many co-workers, in particular with: Jaak Berendson, Gunnar Bergman, Göran Camitz, Kaija Eistrat, Göran Engström, Mats Foghelin, Jan Gullman, Sven Hedman, Rolf Holm, Torsten Johnsson, Lage Knutsson, Vladimir Kucera, Christofer Leygraf, Mats Linder, Bo Mannerskog, Sune Ström, Bengt Sundvall, and Gösta Svendenius. Ove Nygren has taken the majority of the photographs and Bengt Sundvall has made the drawings. The translation into English has been made by Brian Hayes. These valuable contributions are gratefully acknowledged.

I also wish to thank warmly my friend Tony Mercer, who has checked the manuscript from linguistic point of view and also suggested many technical improvements.

Finally, I hope that this book will be of interest and value to the readers and effectively help to combat corrosion.

Stockholm
31 October 1988

Einar Mattsson

1
Corrosion and its importance to the community

The word *corrosion* comes from the Latin word 'corrodere', which means 'gnaw away' . The rusting of iron and steel is the most well-known form of corrosion. Similar processes occur in other metals and also in non-metallic materials, such as plastics, concrete and ceramics. According to the definition, the term 'corrosion' stands for a process. This takes place via a physicochemical reaction between the material and its environment and leads to changes in the properties of the material. The result is a *corrosion effect* which is generally detrimental but can sometimes be useful. Examples of detrimental corrosion effects are *corrosion attack* on the actual material, contamination of the environment with corrosion products and functional impairment of the system, e.g. a steam power plant, in which both the material and the environment form parts. The disintegration of scrap metal such as empty cans and abandoned cars are examples of useful corrosion effects (Fig. 1), as is the deliberate use of corrosion processes in, for example, the reaction between steel and phosphoric acid to produce a phosphated surface suitable for painting. Generally, however, corrosion is a deleterious process and causes a great deal of destruction and inconvenience in our communities. Some examples are as follows:

— *The operational reliability of structures can be jeopardised.* This is the case in, for example, underground water mains, which can be put out of action by corrosion (Fig. 2). Other examples occur in electronic equipment where important control functions can be affected by corrosion, on offshore oil platforms which operate under extremely corrosive conditions and in nuclear power stations where corrosion damage can give rise to costly production breakdowns which in certain cases are completely unacceptable from the safety viewpoint.

— *Natural resources are lost.* Eventually corrosion leads to energy losses, since energy is consumed during the production of metals from their ores, and the corrosion process returns the metal to the ore. Furthermore, corrosion products, which may be widely distributed by water and wind cannot be used to reclaim metal.

— *The environment can be damaged.* Underground oil tanks, which become perforated as a result of corrosion, are an example where a threat is posed to ground water (Fig. 3).

Fig. 1 A scrap car left to be 'destroyed' by the environment.

Fig. 2 Corrosion damage to mains water pipe which caused about 10,000 households to be without water.

Many attempts have been made to estimate the costs to the community caused by corrosion. These include the costs which can arise in the form of corrosion protection measures, through replacement of corrosion-damaged parts or through different effects deriving from corrosion, such as shut-down of production or accidents which lead to injuries or damage to property. Several estimations have arrived at the conclusion that the total annual corrosion costs in the industrialised countries amount to about 4% of the gross national product. Parts of these costs are unavoidable since it would not be economically viable to carry out the necessary precautions to eliminate completely corrosion damage. It is, however, certain that one could reduce losses considerably solely by better exploiting the knowledge we have today and, according to one estimate, about 15% of corrosion costs are of this type [1].

Fig. 3 Pitting in an oil tank after 6 years underground (Bergsöe & Son).

With regard to the importance to the community of corrosion it is vital that every engineer during his education is made aware of the effects and implications of corrosion and that the knowledge available is stored in such a way that it is easily retrievable and ready to be used.

Technical advances continuously bring in their train, however, new corrosion problems. Thus, on the one hand, the corrosion behaviour of new materials needs to be evaluated, while, on the other, well-known materials may be used in new situations so that new corrosion environments are created. For this reason, the current level of knowledge is not sufficient and further research and development are required in the area of corrosion, to complement technical developments at large.

2
Basic electrochemical concepts

The processes involved in corrosion are to a large extent electrochemical in nature. Before venturing into the deeper discussion of corrosion a summary will be given of basic electrochemical concepts.

2.1 ELECTROCHEMICAL REACTIONS

An *electrochemical reaction* is characterised by the fact that it takes place with donating or receiving of electrons. Such a reaction can be schematically represented:

$$\mathrm{Ox} + ne^- \underset{\text{oxidation}}{\overset{\text{reduction}}{\rightleftharpoons}} \mathrm{Red}\ ,$$

where Red is a *reducing agent* (electron donor), Ox is an *oxidising agent* (electron acceptor) and n is the number of electrons (e^-), taking part in the reaction.

When a reaction takes place with the emission of electrons, i.e. when it goes to the left, we refer to it as *oxidation.* If it takes place with consumption of electrons, i.e. it goes to the right, then it is *reduction*. A reducing agent and an oxidising agent associated as indicated by the formula are often called a *redox pair* and the reaction a *redox reaction.*

Electrons cannot exist free in a solution in any significant concentration. The electrons emitted in an oxidation reaction must therefore be used up via a simultaneous reduction reaction.

A possibility is that the reactions take place by contact between an oxidising agent and a reducing agent, both of which are in solution, for example:

oxidation: $2Fe^{2+} \rightarrow 2Fe^{3+} + 2e^-$

reduction: $\frac{1}{2}O_2 + H_2O + 2e^- \rightarrow 2OH^-$

---

total reaction: $2Fe^{2+} + \frac{1}{2}O_2 + H_2O \rightarrow 2Fe^{3+} + 2OH^-$.

An alternative is that the reaction takes place in an *electrochemical cell* (Fig. 4). An electrochemical cell of the ordinary kind consists of two *electrodes* connected by an *electrolyte*. The electrodes are made of an electron conductor, e.g. a metal which is in contact with an electrolyte. The electrolyte is often an aqueous solution having the ability to conduct electricity. A characteristic is that the current through the electrolyte is transported by ions. The electrode from which the positive electric current enters the electrolyte is called the *anode*. The other electrode, through which the electric current leaves the electrolyte, is called the *cathode*.

When the electric current passes through an electrode surface in one direction or the other, an electrochemical reaction always takes place. This is called an *electrode reaction*. The electrode reaction at the anode, i.e. the *anode reaction*, is always an oxidation reaction. The *cathode reaction* on the other hand is always a reduction reaction.

An electrochemical cell in which the current flow is forced by an external source, is called an *electrolytic cell*. An electrochemical cell which can itself produce an electric current is called a *galvanic cell*.

2.2 FARADAY'S LAW

The current entry into and exit from an electrolyte is consequently always associated with electrode reactions which will be manifested as changes in the electrode materials or the environment. The quantities converted during electrode reactions are proportional to the amount of current which passes through the electrode surface. This is expressed in *Faraday's law*, according to which it requires 96 500 coulombs (ampere-seconds) or 26.8 Ah (ampere-hours) for a conversion including one mole of electrons (e^-).

The *current efficiency* for a given electrode reaction gives the proportion of current through the electrode surface, which is required for that reaction. The remainder of the current is consumed by other electrode reactions which take place simultaneously at the electrode surface.

2.3 THE CONCEPT OF ELECTRODE POTENTIAL

If a piece of metal, Me, is surrounded by an aqueous solution containing ions of the metal, Me^{n+}, then electrode reactions take place at the surface of the metal until equilibrium is reached:

$$Me^{n+} + ne^- \rightleftharpoons Me \ .$$

These reactions lead as a rule to an *electrical double layer* being established in the interface region (Fig. 5).

The existence of this electrical double layer means that the piece of metal now has another electrical potential, that is, the so-called Galvani potential (ϕ_1) than the solution (ϕ_2).

The difference in Galvani potential, $\phi_1 - \phi_2$, cannot as a rule be determined by direct measurement but a relative value can be measured in comparison with the Galvani potential difference of a so-called reference electrode (see 2.5). This measurable relative value is called the *electrode potential* and is designated by the symbol E.

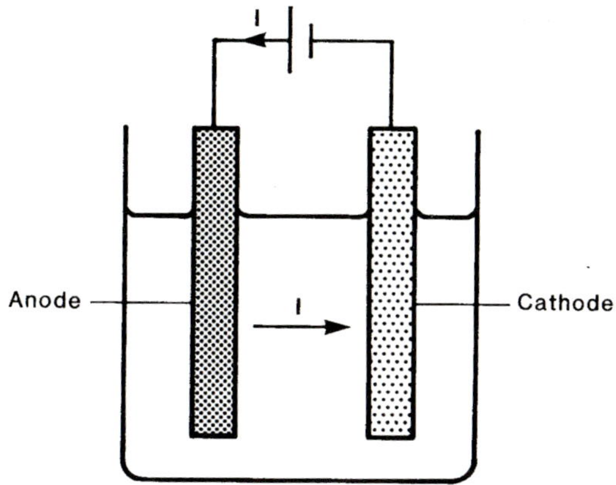

Fig. 4 Electrochemical cell; the arrows show the direction of the electric current (I).

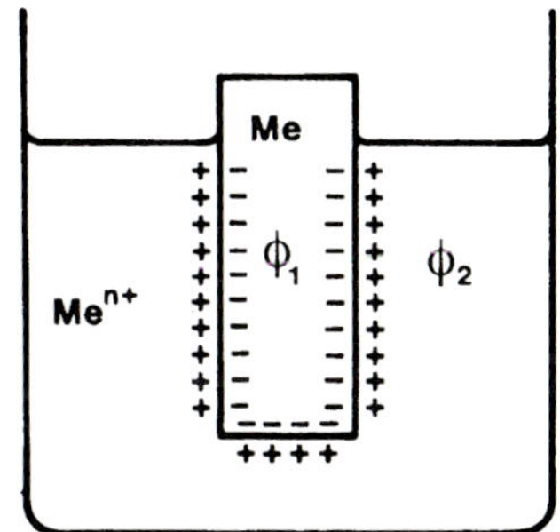

Fig. 5 Metal (Me) in an aqueous solution containing the metal ions, Me^{n+}; an electrical double layer has formed.

At equilibrium a relationship exists between the electrode potential (*equilibrium potential*) and the activity of the metal ion, i.e. the 'effective' metal ion concentration in the solution. This relationship is given by the *Nernst equation*:

$$E = E^0 + \frac{0.0001983 \cdot T}{n} \cdot \log_{10} a_{Me^{n+}} ,$$

where E^0 *is the standard electrode potential*, a constant characteristic for the electrode reaction, T is the absolute temperature and $a_{Me^{n+}}$ is the metal ion activity, i.e. the 'effective' metal ion concentration, which can often be replaced by the metal ion concentration for approximate calculations.

Similar conditions apply, when an inert metal (non-reacting metal such as platinum or gold) is surrounded by a solution containing a redox pair. Reactions take place in accordance with the following formula, until equilibrium has been achieved:

$$\mathrm{Ox} + ne^- \underset{\text{ox}}{\overset{\text{red}}{\rightleftharpoons}} \mathrm{Red}\ .$$

Even these reactions lead to an electrical double layer being formed in the surface zone, where the inert metal is in contact with the solution (see Fig. 5). The electrode potential in this case is called the *redox potential* of the solution. A high redox potential means that the solution is highly oxidising.

Analogous to the relationship between a metal in contact with a solution containing ions of the metal, is the relationship between the redox potential and the activities of Ox and Red. This relationship is also given by the Nernst equation, which is then written:

$$E = E^0 + \frac{0.0001983 \cdot T}{n} \cdot \log_{10} \frac{a_{\mathrm{Ox}}}{a_{\mathrm{Red}}}\ .$$

In actual fact the Nernst equation for $\mathrm{Me/Me}^{n+}$ can be regarded as a special case of this equation, where $a_{\mathrm{Red}} = a_{\mathrm{Me}} = 1$.

2.4 THE GALVANIC CELL

It was mentioned in section 2.1 that an electrochemical cell having the ability to produce electric current itself, is called a galvanic cell (Fig. 6).

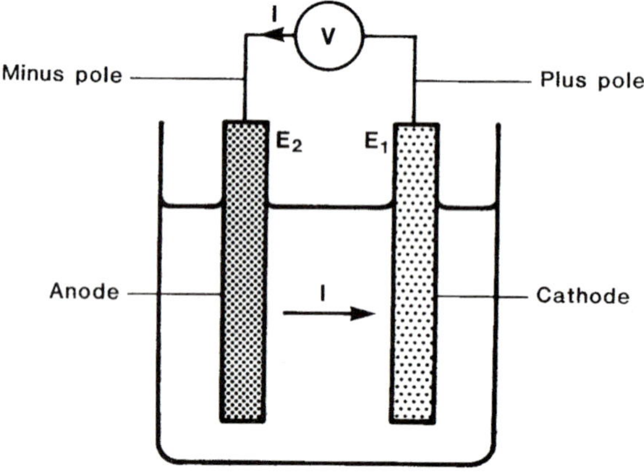

Fig. 6 Galvanic cell with voltmeter (V).

If the electrodes of such a galvanic cell are coupled together via the external metal conductor, an electric current will flow from one electrode (positive pole) to the other (negative pole). † On the other hand, in the electrolyte the current will flow in the opposite direction, as indicated in Fig. 6. The positive pole acts therefore as the cathode and the negative pole as the anode.

The electrical potential difference between the electrodes, which can be measured with a voltmeter is called the *terminal voltage*. The terminal voltage measured when the galvanic cell is not producing current is called the *electromotive force* or *e.m.f.* The electromotive force is a measure of the driving force of the chemical reaction that will take place in the cell, when the electrodes are connected by a metal conductor. The electromotive force (ΔE) can be calculated from the electrode potentials (E_1 and E_2) of the electrodes making up the cell:

$$\Delta E = E_1 - E_2.$$

One can differentiate between different kinds of galvanic cell:

— *bimetallic cells*, (which have in the past been called galvanic cells), where the electrodes consist of different materials,

— *concentration cells* where the electrodes consist of the same material but where the concentrations (activities) of the substances taking part in the reactions are different at the electrodes, and

— *thermo-galvanic cells*, where the electrodes consist of the same material and where the composition of the electrolyte within the cell is constant but where the temperatures at the electrodes are different.

Each electrode with its surrounding electrolyte is called a *half cell*. It can be seen from the above that the two half cells of a galvanic cell can be in contact with the same electrolyte or with different electrolytes. In the latter case, the electrolytes may be separated with the aid of a membrane, which could, for example, have the characteristic of allowing ion exchange, i.e. current flow. In some cases the two half cells can be joined by a liquid junction, as shown in Fig. 7. The liquid junction

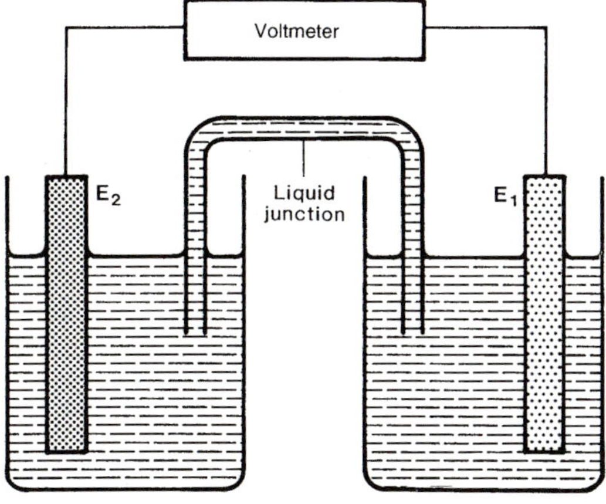

Fig. 7 Half cells connected by a liquid junction.

† In certain instances the electron flow is shown and this will be in the opposite direction.

consists of electrolyte possibly absorbed in a thickening agent, e.g. a gel of agar-agar. A galvanic cell having a liquid junction is written schematically in accordance with international practice as shown in the following example, where the positive pole is placed on the right:

$$Zn \,|\, Zn^{2+} \,||\, Cu^{2+} \,|\, Cu \ .$$

2.5 REFERENCE ELECTRODES

The electrode potential of an electrode has, as mentioned in section 2.3, an absolute value, the Galvani potential difference, but this is difficult to determine experimentally. As a rule one must be satisfied with a relative value. The *test electrode*, i.e. the electrode whose electrode potential is to be determined, is connected via an electrolyte, a liquid junction, to a so-called *reference electrode* (Fig. 8). This consists

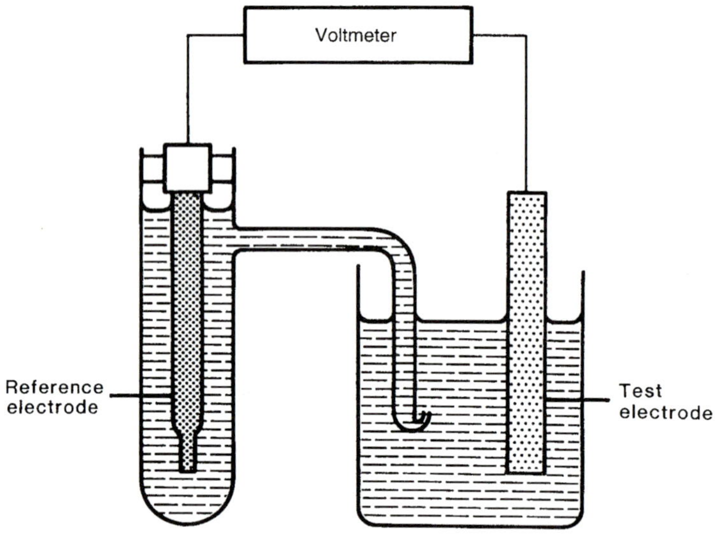

Fig. 8 Determination of electrode potential with the aid of a reference electrode.

of a half cell, which is characterised by a constant and reproducible electrode potential.

With the aid of a high impedance voltmeter the electromotive force (ΔE) of the electrochemical cell thus formed, is measured. The measurement is carried out in such a way that the flow of current is as low as possible:

$$E = E_{test} - E_{ref}$$
$$E_{test} = \Delta E + E_{ref} \ .$$

The absolute value of the electrode potential of the test electrode is therefore the sum of the measured electromotive force ΔE and a constant which is equal to the electrode potential of the reference electrode. Because it is very difficult to determine the latter, one is restricted, as mentioned above, to quoting the potential of the working electrode in comparison with that of the reference electrode, i.e. by the measured value, ΔE. In order to give this value unambiguously one must specify the reference electrode to which the value is referred.

Examples of reference electrodes are:

— *the standard hydrogen electrode* (SHE) consisting of a platinum wire platinised by electrolysis, which is surrounded by a solution having a H^+ ion activity of 1 and bathed in hydrogen gas at 1 atm pressure,
— *the calomel electrode*, which consists of mercury in contact with mercury (I) chloride (calomel) and potassium chloride solution of a given concentration, e.g. 0.1 M, 1 M or saturated solution (SCE),
— *copper/copper sulphate electrode*, which consists of copper in contact with a saturated copper sulphate solution ($Cu/CuSO_4$).

The convention is that the standard hydrogen electrode is given the value zero. Electrode potentials, referred to this zero point, are said to be given on the *hydrogen scale* and are designated E_H. In Table 1 the electrode potentials (on the hydrogen scale) for some of the most common reference electrodes are listed. In technical and experimental work it is not usual to make measurements directly against a standard hydrogen electrode. However, knowing the electrode potential on the hydrogen scale of another reference electrode which is more convenient to use in practice, the measured value of electrode potential can easily be transformed to a value on the hydrogen scale. The electrode potentials given in relation to the saturated calomel electrode are often designated E_{SCE}.

Table 1 Electrode potentials for some common reference electrodes on the hydrogen scale, 25°C

	Designation	E_H (V)
Standard hydrogen electrode	H_2 (1 atm) $\vert$ H^+ (a = 1)	0
Calomel electrode (saturated)	Hg $\vert$ Hg_2Cl_2, KCl (saturated)	+ 0.244
Calomel electrode (1.M)	Hg $\vert$ Hg_2Cl_2, KCl (1 M)	+ 0.283
Calomel electrode (0.1M)	Hg $\vert$ Hg_2Cl_2, KCl (0.1 M)	+ 0.336
Silver/silver chloride electrode (0.1 M)	Ag $\vert$ AgCl, KCl (0.1 M)	+ 0.288
Copper/copper sulphate electrode (saturated)	Cu $\vert$ $CuSO_4$ (saturated)	+ 0.318
Manganese dioxide electrode	MnO_2 $\vert$ Mn_2O_3, NaOH(0.5 M)	+ 0.405

2.6 THE ELECTROCHEMICAL SERIES

Every electrode reaction has its standard potential (see 2.3). This is the electrode potential that occurs when all of the substances taking part in the electrode reaction have activity = 1. If the electrode reactions are arranged according to the values of standard potential, then the *electrochemical series* (Table 2) is obtained. A metal

Table 2 The electrochemical series at 25°C

Electrode reaction	Standard electrode potential, E_H (V)
$Au^{3+} + 3e^- \rightleftharpoons Au$	+1.42
$Cr_2O_7^{2-} + 14H^+ + 6e^- \rightleftharpoons 2Cr^{3+} + 7H_2O$	+1.36
$Cl_2 + 2e^- \rightleftharpoons 2Cl^-$	+1.36
$O_2 + 4H^+ + 4e^- \rightleftharpoons 2H_2O$	+1.23
$Ag^+ + e^- \rightleftharpoons Ag$	+0.80
$Cu^+ + e^- \rightleftharpoons Cu$	+0.52
$Cu^{2+} + 2e^- \rightleftharpoons Cu$	+0.34
$H^+ + e^- \rightleftharpoons \frac{1}{2}H_2$	0
$Pb^{2+} + 2e^- \rightleftharpoons Pb$	−0.13
$Sn^{2+} + 2e^- \rightleftharpoons Sn$	−0.14
$Ni^{2+} + 2e^- \rightleftharpoons Ni$	−0.23
$Co^{2+} + 2e^- \rightleftharpoons Co$	−0.28
$Cd^{2+} + 2e^- \rightleftharpoons Cd$	−0.40
$Fe^{2+} + 2e^- \rightleftharpoons Fe$	−0.41
$Cr^{3+} + 3e^- \rightleftharpoons Cr$	−0.74
$Zn^{2+} + 2e^- \rightleftharpoons Zn$	−0.76
$Mn^{2+} + 2e^- \rightleftharpoons Mn$	−1.03
$Ti^{2+} + 2e^- \rightleftharpoons Ti$	−1.63
$Al^{3+} + 3e^- \rightleftharpoons Al$	−1.71
$Mg^{2+} + 2e^- \rightleftharpoons Mg$	−2.38
$Na^+ + e^- \rightleftharpoons Na$	−2.71
$Ca^{2+} + 2e^- \rightleftharpoons Ca$	−2.76
$K^+ + e^- \rightleftharpoons K$	−2.92
$Li^+ + e^- \rightleftharpoons Li$	−3.05

which corresponds to a relatively high standard potential, e.g. copper, is said to be a *noble metal*. A metal which, on the other hand, corresponds to a low standard potential, e.g. sodium or magnesium, is said to be a *base metal*.

It should be noted that the electrochemical series applies only to oxide-free metal surfaces and at the activities (concentrations), for which the standard potentials are valid. In actual practice the metal surfaces are often covered by oxide films. Furthermore, the activities can deviate considerably from 1, especially when the metal ions are associated with other constituents in so-called complex ions. Such conditions can result in the measured potentials having a completely different order than that given in the electrochemical series. If metals exposed to a given electrolyte, e.g. sea water, are arranged according to measured electrode potentials, then a *galvanic series* is produced. This is valid, however, only for the electrolyte in question, i.e., in which the measurements were made. In Table 3 the galvanic series is given for some pure metals and alloys in sea water at 20°C.

Table 3 A galvanic series of some metals in sea water at 20°C

↑ More noble metals
↓ Less noble metals

Metal	Electrode potential, E_H (V)
Gold	+0.42
Silver	+0.19
Stainless steel (18/8) in the passive state*	+0.09
Copper	+0.02
Tin	−0.26
Stainless steel (18/8) in the active state*	−0.29
Lead	−0.31
Steel	−0.46
Cadmium	−0.49
Aluminium	−0.51
Galvanised steel	−0.81
Zinc	−0.86
Magnesium	−1.36

* In the passive state the metal surface has a thin passivating coating, which does not occur in the active state (see 8.2).

2.7 THE PHENOMENON OF POLARISATION

When a metal is exposed to an aqueous solution containing ion of that metal, there occur at the surface of the metal, both an oxidation of the metal atoms to metal ions, and a reduction of metal ions to metal atoms according to the formula:

$$Me^{n+} + ne^{-} \rightleftharpoons Me\ .$$

Because an exchange of electrons takes place the rate of the two reactions can be given by two different current densities (current density = current strength per unit of surface area) $\overleftarrow{i}$ and $\overrightarrow{i}$. At equilibrium (E_0) $\overleftarrow{i} = \overrightarrow{i} = i_0$, assuming no other electrode reactions take place at the electrode surface, i_0 being called the *exchange current density*. The electrode potential, the so-called equilibrium potential, can thus be calculated according to the Nernst equation (see 2.3).

Applying a net current, i, to the surface means that $\overleftarrow{i} \neq \overrightarrow{i}$. The applied net current density will in actual fact be the difference between $\overleftarrow{i}$ and $\overrightarrow{i}$. When current is applied to the electrode surface the electrode potential is changed and takes on the value E_i. The electrode is said to be *polarised*. The change in electrode potential is called *polarisation* and is usually designated by the greek letter eta, η:

$$\eta = E_i - E_0 .$$

The polarisation (corrected for possible errors in measurement) can be divided into two main components:

— *Concentration polarisation*, caused by the difference in concentration between the layer of electrolyte nearest the electrode surface, i.e. the diffusion boundary layer, and the bulk of electrolyte (see 2.8).

— *Activation polarisation*, caused by a retardation of the electrode reaction.

The polarisation of an anode is always positive and that of a cathode always negative (Fig. 9).

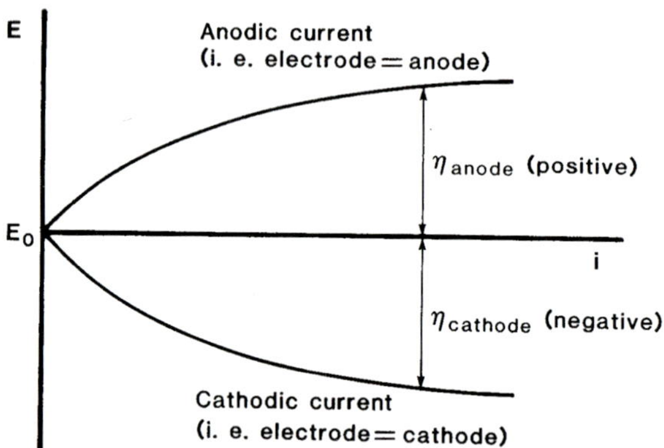

Fig. 9 Anodic and cathodic polarisation curves; E = electrode potential, i = current density and η = polarisation.

Polarisation thus reduces the terminal voltage of a galvanic cell, as the current drawn from the cell increases. When current is passed through an electrolytic cell from an external source the polarisation produced will lead to the need for a higher applied voltage. In a case where polarisation can be assigned to a definite electrode reaction, then one can refer to it as *overpotential*. Overpotential is the difference between the electrode potential of the electrode with applied current and the equilibrium potential for the electrode reaction in question. Hydrogen over potential, for example, occurs on electrolytic evolution of hydrogen according to:

$$H^+ + e^- \rightarrow \tfrac{1}{2}H_2 ,$$

and oxygen overpotential occurs on the production of oxygen by electrolysis according to:

$$2OH^- \rightarrow \tfrac{1}{2}O_2 + H_2O + 2e^-.$$

However, it often is the case that two or more electrode reactions can take place simultaneously at an electrode, e.g. a cathode reaction, whose equilibrium potential

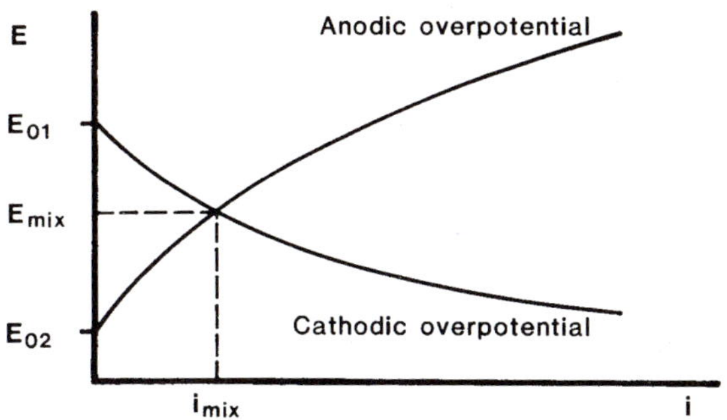

Fig. 10 Overpotential curves for two electrode reactions, occurring simultaneously at an electrode surface; E = electrode potential, i = current density.

is E_{01}, and an anode reaction, whose equilibrium potential is E_{02} (Fig. 10). Since there is no net current applied to the electrode surface, the anodic and cathodic current densities are alike (i_{mix}) and the electrode shows a so-called *mixed potential* (E_{mix}), equivalent to the point of intersection between the anodic and the cathodic overpotential curves. The change in electrode potential which takes place when current is applied to such an electrode is also to be regarded as polarisation.

Activation polarisation at low polarisation values is directly proportional to the current density. At higher polarisation values (> *ca* 30-50 mV) there is, on the other hand, a linear relationship between activation polarisation and the logarithm of the current density. This relationship is given by the *Tafel equation*:

$$|\eta| = a + b \cdot \log_{10} i \ .$$

Polarisation curves are therefore drawn, as a rule, as a function of $\log_{10} i$, so that straight lines are produced, i.e., the so-called *Tafel lines*. At low current densities the concentration polarisation is often negligible, but it can be dominating at high current densities (Fig. 11).

When recording conventional polarisation curves direct current is applied and steady state conditions are awaited. The electrode is then equivalent to a resistance, the polarisation resistance, which, however, varies with current density. But when alternating current is applied, the electrode appears to be equivalent to a more

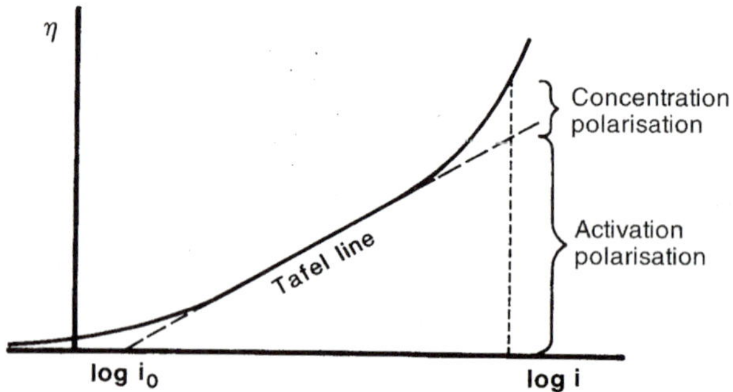

Fig. 11 Polarisation (η) as a function of logarithm of current density (i); the continuous line indicates the total polarisation, i_0 = the exchange current density.

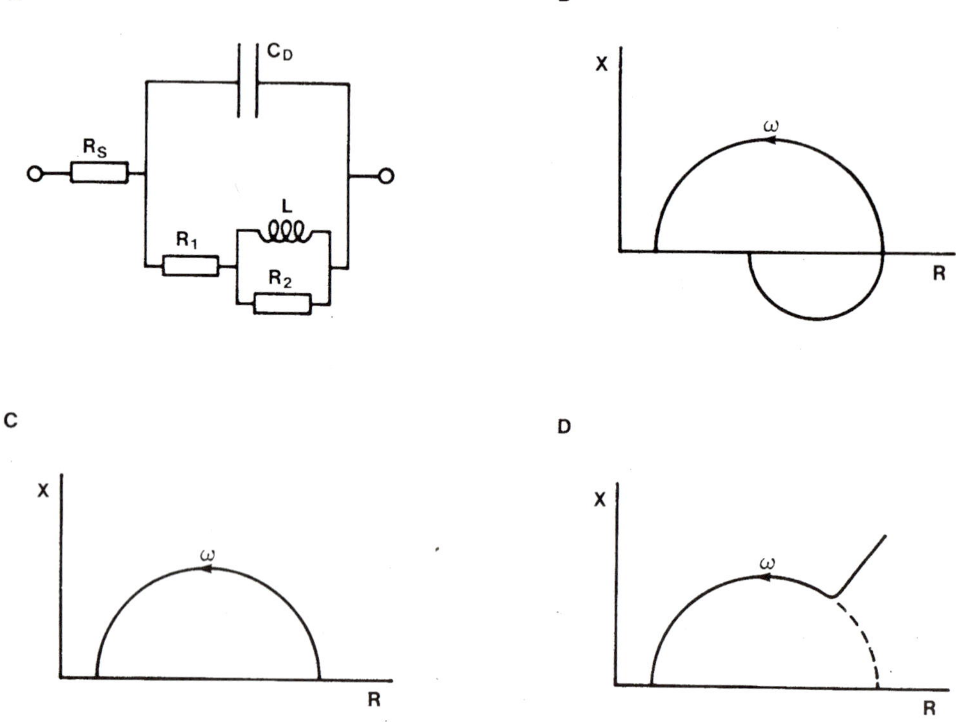

Fig. 12 A. Example of electrical analogue of an electrode; R = electrolyte resistance, C_D = capacitance of double layer, L = inductance, R_1 and R_2 = resistances. B–D. Impedance spectra for some different types of electrodes; R = resistive part, X = reactive part; the arrows show increased frequency (ω).

complicated electrical model having a polarisation impedance, containing resistive as well as capacitive and inductive components. In Fig. 12A an example is given of such an electrical model. By measuring the (AC) impedance of alternating current an electrode one can obtain information about the electrode processes over and above those obtainable by measurement of direct current polarisation. This technique is called *electrochemical impedance spectroscopy (EIS)*. In Fig. 12 B–D examples are given of impedance spectra obtained by varying the frequency of the alternating current (ω) for different types of electrodes. The actual form of the curve is decided by the internal relations between the different components of electrode impedance and reveals the rate determining step(s) of the electrode process.

2.8 ELECTROLYTIC CONDUCTANCE

The transport of current in an electrolyte solution takes place via the movement of *ions*; *anions* (negatively charged) and *cations* (positively charged) (Fig. 13). Cations

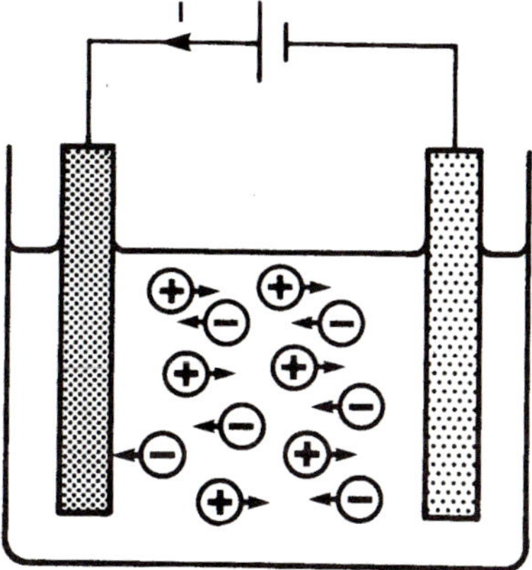

Fig. 13 Transport of electric current through electrolyte.

migrate with the current and anions against it. In the different parts of the solution there is always electro-neutrality which means that the anions and cations balance each other.

Ohm's law applies for transportation of current through the electrolyte solution:

$$U = IR,$$

where I is the current strength in amperes, U is the potential difference between the electrodes in volts and R is the resistance of the electrolyte in ohms.

Further:

$$\frac{1}{R} = L \,,$$

where L is the conductance of the electrolyte in ohm^{-1}.

For a liquid volume with rectangular surfaces:

$$L = \kappa \frac{q}{l} = \frac{q}{\rho l} \,,$$

where q is the cross-sectional area of liquid in cm^2, l is the length of liquid in cm, κ is the specific conductance or conductivity in $ohm^{-1}\,cm^{-1}$, and ρ is the specific resistance or resistivity in ohm cm.

Different ions contribute to differing degrees in transporting the current through the electrolyte solution. Thus, one refers to the ions having different mobilities or *ionic conductivities*. The part of the current carried by a certain ionic species in an electrolyte solution is called the *transport number* of that ion.

In spite of the fact that current transport is distributed so differently between the different ions in solution, the current in most cases does not lead to any significant concentration changes in the bulk of the electrolyte. This is because equalisation takes place relatively quickly via liquid movement (convection). However, in a thin solution film (10^{-1}–10^{-5}cm thick) close to the electrode surfaces, i.e. in a so-called *diffusion boundary layer*, where convection is difficult, current transport can lead to considerable changes in concentration. This can mean that, for example, the pH value near to the electrode surface will be considerably different from that in the bulk of solution. Another consequence is concentration polarisation (see 2.7).

3
Basic corrosion concepts

Within corrosion theory one can distinguish certain concepts which are of fundamental importance. Furthermore, one can divide the subject area in various ways, e.g. according to the corrosion type, corrosive medium, the method of corrosion protection, the type of material or field of application etc. Each basis for division provides its own possibilities for emphasising special features of corrosion phenomena. In this connection the importance of standardised language must be emphasised. The nomenclature which has been used in this book is in accordance with the international standard ISO 8044. According to this terminology three terms are of fundamental importance: corrosion, corrosion effect and corrosion damage (Table 4).

Table 4 Definitions of some basic corrosion concepts according to the international standard ISO 8044

Corrosion	Physicochemical interaction between a metal and its environment which results in changes in the properties of the metal and which may often lead to impairment of the function of the metal, the environment, or the technical system of which these form a part. *Note* — This interaction is usually of an electrochemical nature.
Corrosion effect	Change in any part of the corrosion system caused by corrosion.
Corrosion damage	Corrosion effect which is considered detrimental to the function of the metal, the environment or the technical system of which these form a part.

Of primary importance within corrosion theory is the question of whether there exists a driving force for corrosion. If there is, then the degree of retardation of this reaction is of major importance and generally decisive for the rate of corrosion.

3.1 THE DRIVING FORCE OF CORROSION

When metals are produced in blast-furnaces, smelters etc., they are transformed from the stable state in the ore to the metallic state, which is not stable under most conditions found in practice. Therefore, in most metals exposed to the weather, there exists a driving force for them to return to stable compounds similar to those found in the ores. In general it is a 'return' to the ore composition which takes place when a metal corrodes. An example is the rusting of steel. In this process the metallic iron is converted to ferrous/ferric compounds such as oxides and hydroxides (rust). These are similar to minerals such as magnetite (Fe_3O_4) or limonite ($Fe_2O_3{\cdot}xH_2O$). Another example is shown in Fig. 14.

Fig. 14 Fragment from aeroplane wheel manufactured from magnesium alloy which has been completely transformed to a crystalline magnesium mineral after 28 years in the Baltic Sea.

A prerequisite that a metal in a given state (M, M^{n+} or M_2O_n) will spontaneously react and be converted to another state is that the reaction takes place with the release of energy. There exists then, a thermodynamic driving force for the reaction. Should the conversion of a metal require a supply of energy, then this means that the conversion cannot take place spontaneously. There does not exist, in other words, a thermodynamic driving force for the reaction. The metal is in this case thermodynamically stable in the conditions prevailing. The state of a metal which is stable in contact with an aqueous solution depends on such factors as the redox potential and pH value of the solution as well as the temperature of the system. In a so-called *potential-pH diagram* one can get an overall view of which states are stable in different potential-pH conditions, i.e. the meaning of a number of thermodynamic stability constants related to the system. As an example Fig. 15 shows the potential-pH diagram for copper in contact with water at 25°C.

The following applies to the different domains in the potential-pH diagram:

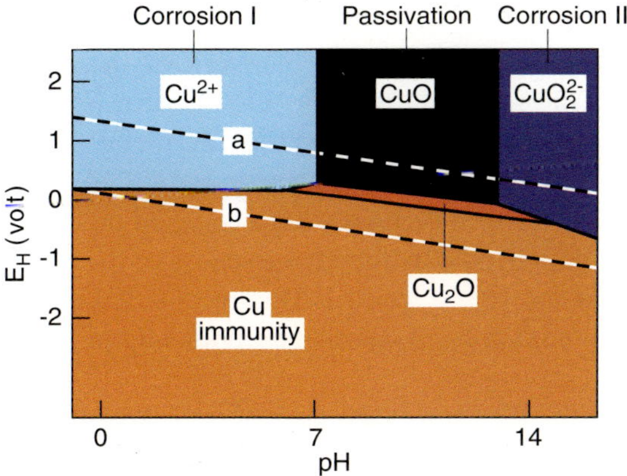

Fig. 15 — Potential-pH diagram for Cu-H_2O at 25°C, 10^{-6} M dissolved Cu; the band between the dashed lines *a* and *b* is the stability domain of water [2].

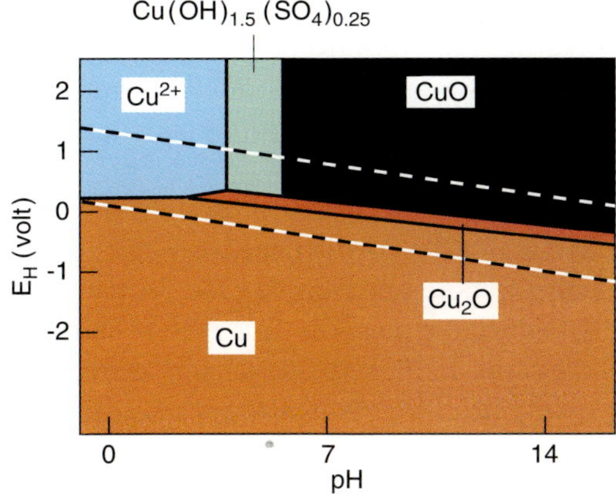

Fig. 16 — Potential-pH diagram for Cu-SO_4^{2-}-H_2O at 25°C, 10^{-3} M SO_4^{2-}, 10^{-1} M dissolved Cu; the domain for $Cu(OH)_{1.5}(SO_4)_{0.25}$ is representative of green patina [3].

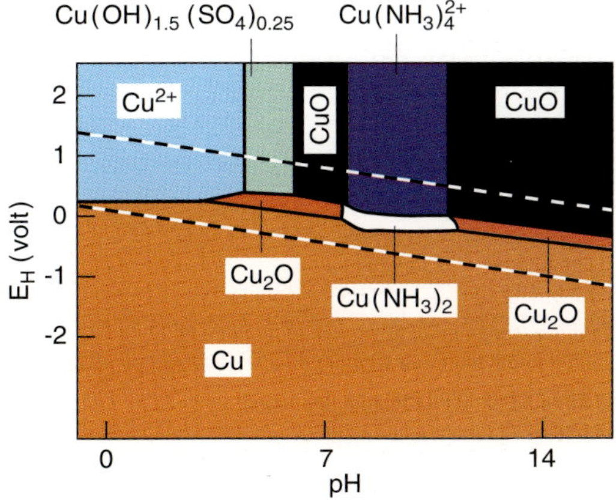

Fig. 17 – Potential-pH diagram for Cu-SO_4^{2-}-NH_3-H_2O at 25°C; 1 M NH_3, 0.5 M SO_4^{2-}, 0.05 M dissolved Cu [4].

Cu^{2+} and CuO_2^{2-}	Soluble species (Cu^{2+} or CuO_2^{2-}) of the metal are stable. There exists therefore, a driving force for *corrosion* of the metal but the diagram provides no information about the rate of corrosion.
CuO and Cu_2O	The solid oxides (Cu_2O or CuO) of the metal are stable. There exists therefore, a driving force for corrosion but the oxide formed can provide a protective coating, a process which is called *passivation*. However, the diagram does not indicate the effectiveness of the protection.
Cu	The metal is the stable species. This means that corrosion is impossible. The metal is said to be *immune* or cathodically protected.

The lines *a* and *b* in the diagram mark the stability domain of water. At electrode potentials above this area, oxidation takes place which leads to the production of oxygen gas. At potentials below the area, reduction takes place with the release of hydrogen gas as a consequence. When a metal is exposed to an aqueous solution, the conditions are as a rule equivalent to a point between the lines *a* and *b*.

The Belgian scientist M. Pourbaix has worked out an atlas with potential-pH diagrams for a large number of systems, most of which are of the type metal–water at 25°C [2]. Potential-pH diagrams are therefore often called *Pourbaix diagrams*. In Fig. 110 and Fig. 129 potential-pH diagrams for iron and aluminium are shown, both in contact with water at 25°C.

In actual practice, however, it is not usual to meet with such simple systems as those given in Figs 15, 110 and 129. This is because water often contains dissolved species, which can cause precipitation or complex ion formation and the presence of such species can radically change the diagram. But even in such cases it is possible to work out a potential-pH diagram. In Figs 16 and 17 it is shown how the potential-pH diagram for copper is changed, when water receives additions of sulphate and sulphate+ammonia respectively.

3.2 ELECTROCHEMICAL AND CHEMICAL CORROSION

Many corrosion processes are, in the widest sense of the expression, electrochemical in nature, because they involve an oxidation reaction:

$$Me \rightarrow Me^{n+} + ne^- \ .$$

In spite of this one can distinguish between:

— *electrochemical corrosion*, which takes place via electrode reactions, usually in a moist environment (corrosion in aqueous solutions is included here, and atmospheric corrosion under the influence of moist films on the surface as well as corrosion in soil aided by moisture), and

— *chemical corrosion*, which takes place under the influence of dry gases (for example high temperature oxidation) or water-free organic liquids.

The major part of corrosion processes is of the electrochemical corrosion type. These processes take place via the action of galvanic cells, which are called *corrosion cells*. There are two main types of such cells (Fig. 18):

— *corrosion cells with separate anode and cathode surfaces*, e.g. aluminium sheet with brass screws, where the aluminium constitutes the anode and the brass the cathode, and

— *corrosion cells without separate anode and cathode surfaces*; the whole metal surface serves as both anode and cathode; the anode and cathode surfaces can be thought of as being small and very numerous, and besides that, can change place at short intervals of time.

In the corrosion cells the following electrode reactions take place:

anode reaction: $\mathrm{Me} \rightarrow \mathrm{Me}^{n+} + ne^-$

cathode reaction: $\mathrm{Ox} + ne^- \rightarrow \mathrm{Red}$

total reaction: $\mathrm{Me} + \mathrm{Ox} \rightarrow \mathrm{Me}^{n+} + \mathrm{Red}$.

In order for electrochemical corrosion of a metal to take place, it is therefore necessary for an oxidising agent to be present, which can be reduced. In most cases of corrosion the role of oxidising agent is taken by oxygen dissolved in water or in some cases and particularly in acid solutions, by H^+ ions. Then the cathode reactions are

$\frac{1}{2}O_2 + H_2O + 2e^- \rightarrow 2OH^-$ and

$H^+ + e^- \rightarrow \frac{1}{2}H_2$ respectively.

3.3 THE RATE OF CORROSION

If a thermodynamic driving force exists for a corrosion process, then it will take place. The rate of corrosion can, however, vary within wide limits. In certain cases it can be large and cause serious damage to the material. In other cases it can be small and of little practical importance; this is a result of inhibition of one or more of the electrode reactions.

The extent of corrosion can be expressed as, for example, the change in mass of the material, the depth of the surface zone which has been corroded away, the number and quantity of pits formed, the amount of corrosion products, changes in the ultimate strength, yield strength or rupture strain of the metal, etc. The changes in the magnitude of these per unit of time is a measure of the corrosion rate. Another measure is in fact the density of the corrosion current for, according to Faraday's law (see 2.2), the amount of metal oxidised at the anode is determined by the anodic current causing the oxidation. In Table 5 some of the most frequently encountered units of corrosion rate are given.

If the corrosion rate is expressed as the corrosion current density then a retardation of the reaction will become apparent in the form of polarisation. The polarisation curves displaying this effect for the electrode reactions in the corrosion process can be shown in a so-called *Evans diagram* (Fig. 19).

In the Evans diagram the polarisation curves have been drawn for both the anodic metal oxidation, and for the cathodic reduction process. The intersection of these two

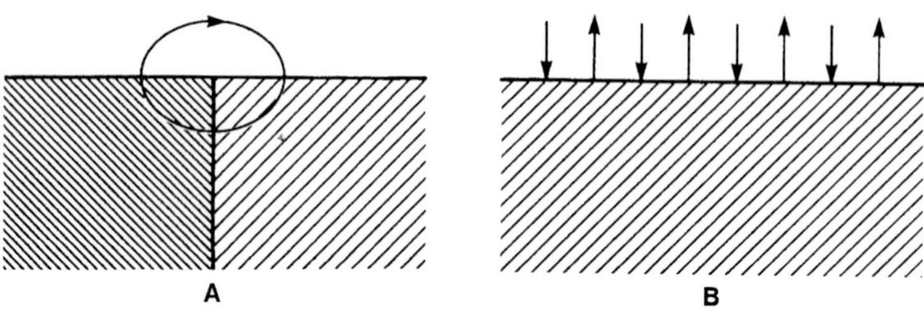

Fig. 18 Corrosion cell. A, anode and cathode surfaces distinguishable. B, anode and cathode surfaces indistinguishable.

Table 5 Various units for corrosion rates

Corrosion effect	Unit
Mass change	g m^{-2}/year
	mg dm^{-2}/day=mdd
Increase in corrosion depth	μm/year
	μm/year=10^{-3} mm/year
	inch per year=ipy=25.4 mm/year
	mil per year=mpy=10^{-3} ipy=25.4 μm/year
Corrosion current	mA cm^{-2}
Decrease in ultimate strength, yield strength or rupture strain	per cent/year (of initial value)

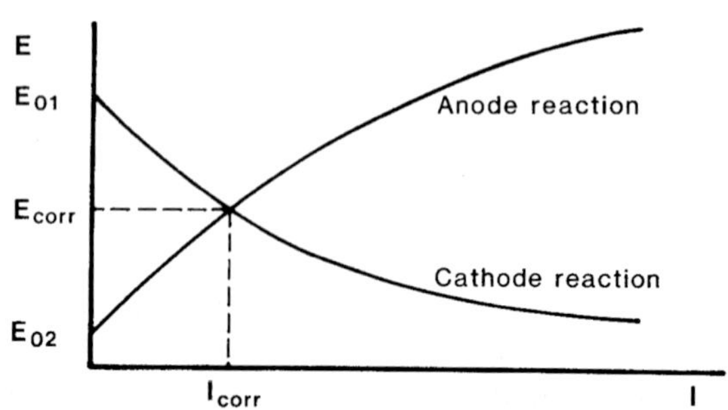

Fig. 19 Evans diagram; *E*=electrode potential, *I*=current.

curves provides information about the corroding electrode. I is the so called *corrosion current* and *E* is the mixed potential, called the *free corrosion potential*, which can be measured for the corroding electrode.

As can be seen from the Evans diagram the position of the intersection and thereby the magnitude of the corrosion current is determined by the form of the polarisation curve. When there is high cathodic polarisation (Fig. 20) the corrosion

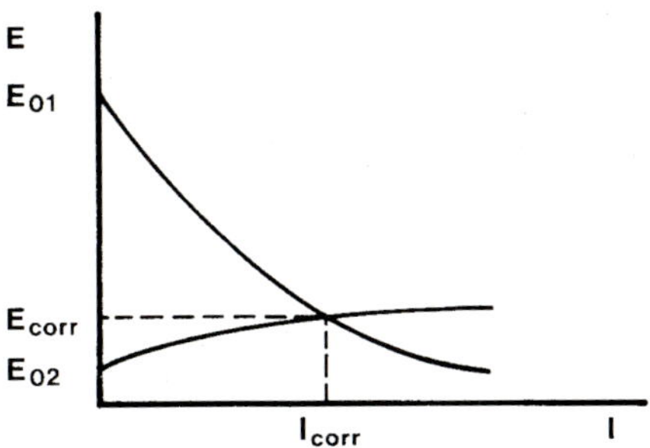

Fig. 20 Cathodic control; *E*=electrode potential, *I*=current.

process is said to be under *cathodic control*. If, on the other hand, the anodic polarisation is dominant (Fig. 21) one talks of *anodic control*. In a case where the

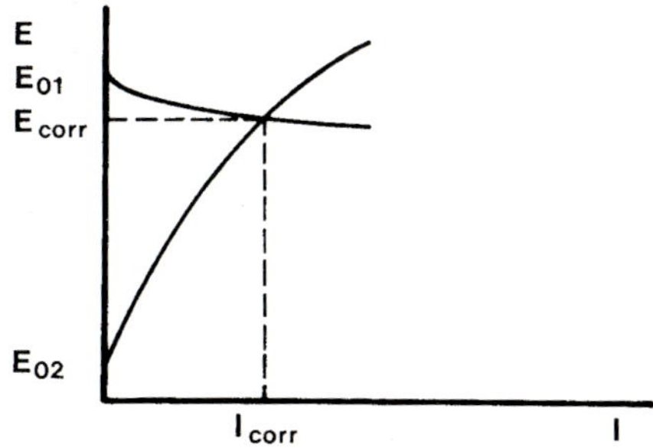

Fig. 21 Anodic control; *E*=electrode potential, *I*=current.

anodic and cathodic polarisations are of the same order of magnitude, the control is *mixed* (Fig. 19). The Evans diagram is applicable in many other similar cases for the schematic description of the polarisation curves of corrosion cells, e.g. to show the influence of inhibitors (see 6.3).

In a *Stern diagram* (Fig. 22) the electrode potential is drawn as a function of

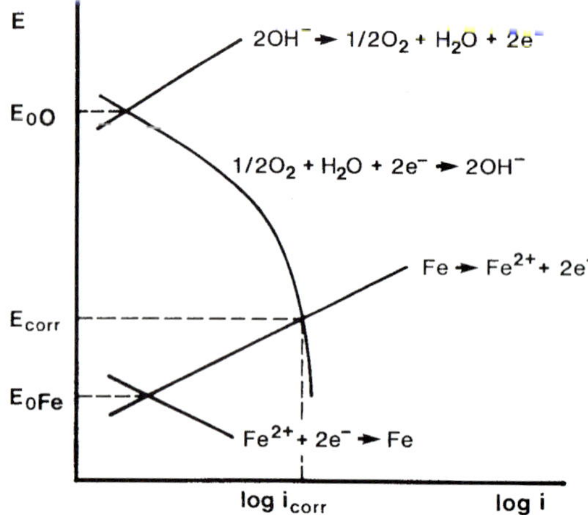

Fig. 22 Stern diagram representative of corrosion of iron in neutral, oxygen-containing solution; the cathode reaction consists of oxygen reduction; E=electrode potential, i=current density.

$\log_{10} i$, and not as a function of I as in the Evans diagram. Here the polarisation curve mostly takes the form of a straight line (Tafel line) within the range of current densities where the concentration polarisation is negligible. In the Stern diagram the equilibrium potentials of the electrode reactions are not situated on the potential axis, as they are in the Evans diagram.

4
Types of corrosion

By taking consideration of the different causes of corrosion and their mechanisms, as well as the appearance of the attack, one can differentiate between several types of corrosion.

4.1 UNIFORM CORROSION

What sets *uniform corrosion* apart is that it proceeds at about the same rate over the whole surface of the metal exposed to the corrosive environment. The extent can be given as mass loss per unit area or by the *average penetration*, which is the average of the corrosion depth. This can be determined by direct measurement or by calculation from the mass loss per unit area, when the density of the material is known. Uniform corrosion takes place as a rule via the effect of corrosion cells without clearly defined anode and cathode surfaces.

With uniform corrosion the usability of the material is usually in direct relation to the average penetration per year. In Table 6 a relationship is given which applies to most cases.

4.2 PITTING

Pitting is localised corrosion which results in pits in the metal surface (Fig. 23). This type of corrosion generally takes place in corrosion cells with clearly separated anode and cathode surfaces. The anode is situated in the pit and the cathode usually on the surrounding surface. Pitting usually results in worse damage than uniform corrosion because it can lead to perforation in a very short period of time.

When evaluating attack caused by pitting the following should be taken into account:

— the number of pits per unit area,
— the diameter of the pits, and
— the depth of the pits.

Table 6 Relations between the usability of materials and their corrosion rates; generally applicable on exposure to liquids when only uniform corrosion occurs

Usability	Average penetration (μm/year)		
	Comparatively expensive materials, e.g. Ag, Ti, Zr	Moderately priced materials, e.g. Cu, Al, stainless steel (18Cr, 9Ni)	Comparatively cheap materials, e.g. carbon steel, cast-iron
Satisfactory	<75	<100	<225
Acceptable under special conditions, e.g. short period of exposure	75–250	100–500	225–1500
Not acceptable without protection	>250	>500	>1500

The number of pits per unit area and the pit diameter are easily determined by comparison with a pictorial standard. In the case of pit depth it is usually the maximum depth that is determined. In some cases this means the depth of the deepest pit observed on the sample examined, but in others the mean value of for example, the five deepest pits. The depth of the pits can be measured with a microscope with focusing first on the bottom of the pit and then on the surface of the uncorroded metal. In this way the distance between the two focusing levels is obtained. The depth of pitting can also be determined with a micrometer or by cutting a cross-section through the pit followed by direct measurement — possibly with the aid of a microscope.

The ratio of the maximum pit depth (P_{max}) and the average penetration (P_{aver}) is called the *pitting factor* (*F*) (Fig. 24):

$$F = \frac{P_{max}}{P_{aver}} \quad .$$

Pitting can occur in most metals. Amongst passivated metals this type of corrosion is initiated only above a certain electrode potential, the *pitting potential.*

4.3 CREVICE CORROSION

Corrosion which is associated with a crevice and which takes place in or immediately around the crevice, is called *crevice corrosion.*

In some cases crevice corrosion can simply be caused by corrosive liquid being held in the crevice, while surrounding surfaces dry out. If the crevice and the surrounding metal surfaces are in a solution, the liquid in the crevice can be almost

Fig. 23 Pits in oil tank of steel.

stagnant. As a result of corrosion in the crevice the conditions there can be changed; for example, the pH value can decrease and the concentration of chloride increase, and as a consequence the corrosivity can be higher in, than outside, the crevice (Fig. 25). When stationary conditions have been established, anodic attack of the metal usually takes place near the mouth of the crevice, while cathodic reduction of oxygen from the surroundings takes place on the metal surfaces outside:

anode reaction: $Me \rightarrow Me^{n+} + ne^-$

cathode reaction: $\frac{1}{2}O_2 + H_2O + 2e^- \rightarrow 2OH^-$.

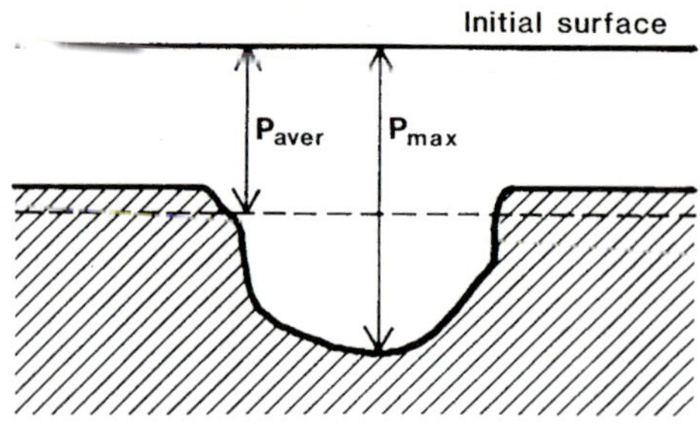

Fig. 24 Pitting factor $F = P_{max}/P_{aver}$.

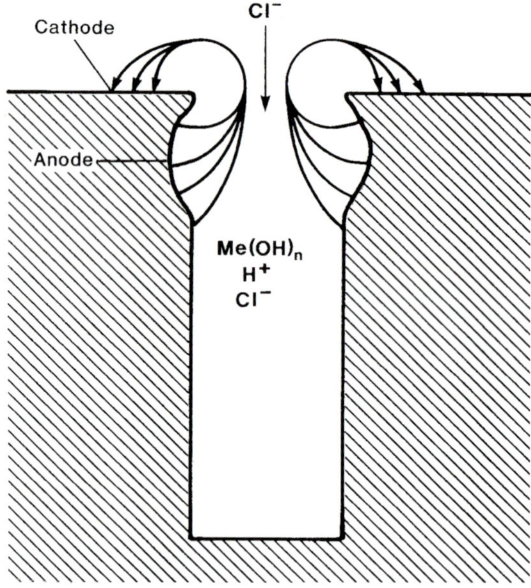

Fig. 25 Principle of crevice corrosion.

Crevice corrosion can take place in most metals. The risk of crevice corrosion should be especially heeded with passive metals, e.g. stainless steel.

Crevice corrosion not only takes place in the crevices between surfaces of the same metal, but also when the metal is touching a non-metallic material. A combination of crevice corrosion and bimetallic corrosion (see 4.12) can take place when two different metals are forming a crevice. The presence of Cl^-, Br^- or I^- ions generally accelerates this type of corrosion. As in the case of pitting, crevice corrosion is initiated only above a certain electrode potential.

Flanged joints of stainless steel in sea water pipes (Fig. 26) can be mentioned as an example of a construction which is at risk for crevice corrosion. A suitable countermeasure is to fill the crevice with durable sealing compound. Another possibility is to use a weld in place of the flanged joint.

4.4 DEPOSIT CORROSION

Corrosion which is associated with a deposit of corrosion products or other substances and which takes place under or immediately around the deposit is called *deposit corrosion.*

This type of corrosion is caused by moisture being held in and under the deposit. Because the movement of water is poor, corrosive conditions can be created under the deposit in a similar way to that described in crevice corrosion. The result is that a corrosion cell is formed with the anode under the deposit and the cathode at, or just outside, the edge (Fig. 27).

Deposit corrosion is found, for example:

— under the road-mud in the wheel arch of a car (Fig. 28),
— under leaves which have collected in guttering, and
— under fouling on ships' hulls and in sea water cooled condensers.

A related corrosion process can occur under the mineral wool insulation around central heating pipes, which have become wet from the water leaking from joints or from rainfall during installation (Fig. 29).

4.5 SELECTIVE CORROSION

Selective corrosion (also known as *dealloying*) can be found in alloys and results from the fact that the components of the alloy corrode at different rates.

The most well-known example of selective corrosion is the *dezincification* of brass (see 8.4) (Fig. 30) . On dezincification the zinc is dissolved selectively, while the copper is left as a porous mass having poor structural strength. Similar corrosion processes include the *dealuminisation* of aluminium bronze and the *selective dissolution of tin* in phosphor bronze.

Graphitic corrosion in grey cast-iron provides another example of selective corrosion where the metallic constituents of the iron are removed (Fig. 31) (see 8.1). The remaining graphite allows the object concerned to maintain its shape but its strength and mass are severely reduced.

4.6 INTERGRANULAR CORROSION

Intergranular corrosion means corrosion in or adjacent to the grain boundaries of the metal.

Fig. 26 Attack by crevice corrosion on a stainless steel flange (AISI 316) in a sea water carrying system.

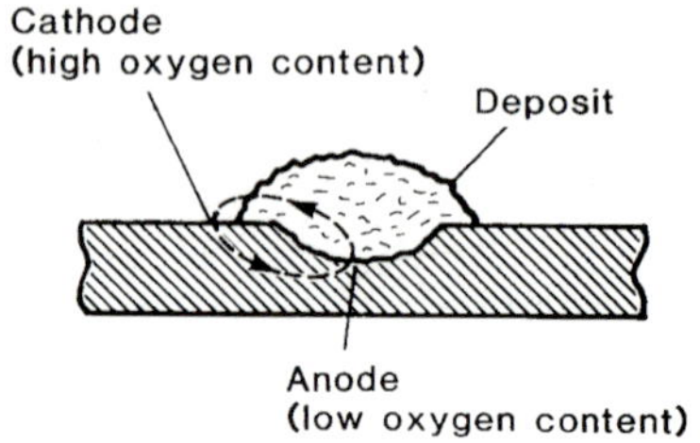

Fig. 27 Principle of deposit corrosion.

Fig. 28 Deposit corrosion in the front wing of a motor car due to road mud accumulation.

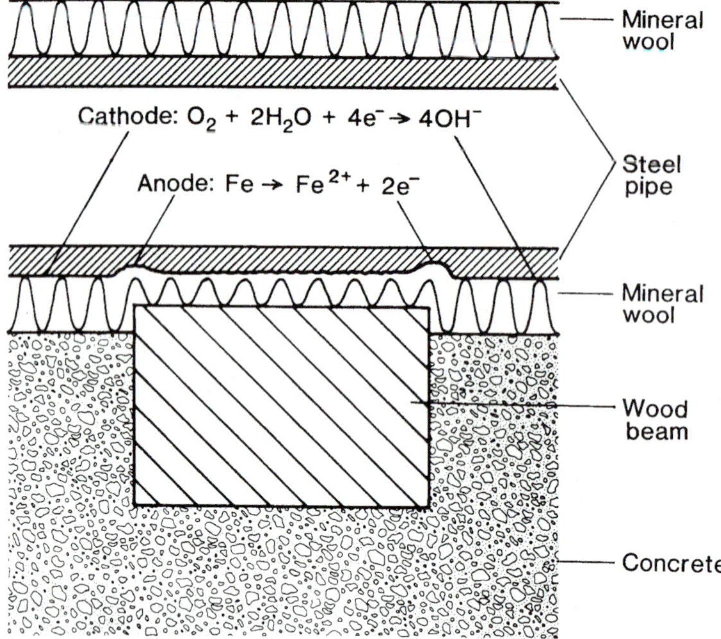

B

Fig. 29 Outside corrosion of a central heating pipe of steel located in a floor structure with damp mineral wool insulation; corrosion is caused by aeration cells at the beam: A, schematic overview; B, attack on the steel pipe.

Metals are usually built of crystal grains. When a metal solidifies or is heat-treated, several processes take place which cause the grain boundary region to take on other corrosion characteristics than the main mass of the grain.

Intergranular corrosion can occur in most metals under unfavourable conditions. Most well-known is intergranular corrosion in stainless steel as a consequence of chromium carbide formation when the carbon concentration is too high and an unfavourable heat-treatment has occurred, e.g. in the heat-affected zone along a weld (Fig. 32) (see 8.2.4).

Fig. 30 Cross-section showing dezincification (reddish zone) of a brass valve cone; the attack has arisen during exposure to sea water (Metallverken).

Fig. 31 Attack by graphitic corrosion in a water valve of cast-iron after 10 months' exposure to hot (50°C) sea water.

4.7 LAYER CORROSION

With *layer corrosion* the attack is localised to internal layers in wrought metal. As a rule the layer is parallel to the direction of processing and usually to the surface. The attack can result in the unattacked layers being detached and looking like the pages in a book (Fig. 33). But the result can also be the formation of blisters swelling the metal surface, because of the voluminous corrosion products. Layer corrosion is rather unusual. It is best known amongst certain aluminium alloys (see 8.3).

Fig. 32 Attack by intergranular corrosion at welds in a stainless steel pipe.

4.8 EROSION CORROSION

Erosion corrosion is a process which involves conjoint erosion and corrosion. It is a type of corrosion usually caused by a rapidly streaming liquid and the attack is thus dependent on the degree of turbulence.

Erosion corrosion can occur with most metals, but copper-bearing materials are particularly sensitive (see 8.4). In many cases the damage appears as a result of the protective coating of corrosion products being damaged in some way or not being able to form due to a high level of turbulence.

Fig. 33 Cross-section showing layer corrosion in aluminium alloy (AlZn5Mg1) in the naturally aged condition; the attack has arisen during 4.5 years' exposure to sea water (Gränges Aluminium) (x30).

Fig. 34 Cross-section showing attack by erosion corrosion in a copper water pipe; direction of flow from left to right (Metallverken) (x 100).

As a rule the pits formed by erosion corrosion have a shiny surface free from corrosion products. They are often undercut in the direction of flow (Fig. 34). Sometimes they have a characteristic horseshoe shape with the shanks pointing in the direction of flow, i.e. the horse is walking up-stream!

Erosion corrosion attack is, as a rule, localised to such areas where the flow is disturbed, e.g. bends in pipes (Fig. 35), the entrance to heat-exchanger pipes, pump parts which are exposed to fast-moving liquids and valve seats in water fittings. Abrasive particles and air bubbles in the liquid increase the risk and intensity of erosion corrosion.

Fig. 35 Attack by erosion corrosion in a bend of a copper water pipe.

4.9 CAVITATION CORROSION

Cavitation corrosion involves conjoint action of corrosion and cavitation. When vapour bubbles formed under reduced pressure collapse, the resulting shock wave can cause material damage. Cavitation corrosion can occur, for example, in rotary pumps, on cylinder liners in engines, and on ships' propellers, especially on fast craft (Fig. 36).

Fig. 36 Attack by cavitation corrosion on a boat propeller of aluminium.

4.10 FRETTING CORROSION

Fretting corrosion is a process involving conjoint corrosion and oscillatory slip between two surfaces in contact. How the reaction proceeds depends on the conditions. Two alternative theories for fretting corrosion of metals in the air may be used to describe the character of the process:

(1) When the metal surfaces glide over each other under pressure, a welding together takes place at certain points. As a result small particles are released with continued movement. These particles of metal quickly oxidise in contact with the

oxygen in the atmosphere to form a finely divided, dark oxide powder.

(2) When in contact with oxygen in the air the metal surface forms an often invisible oxide coating which hinders further oxidation. At the points of contact where the metal surfaces glide over each other there will be a continuous removal and rebuilding of metal oxide. The removed oxide forms a dark powder which discolours the points of contact.

Fretting corrosion can occur in long-distance transportation of stacked or coiled metal sheet and rod material in bundles. The stains are usually streaked longitudinally. Surfaces facing each other show a mirror-imaged pattern of stains, which mark the points of contact (Fig. 37).

Fretting corrosion can also occur in tight-fitting constructions, e.g. in bearings. The examples of fretting corrosion quoted above do not require the presence of water; on the contrary they may be inhibited by water vapour in the atmosphere. Thus, fretting corrosion has caused certain problems in space flights, where the water vapour pressure is low.

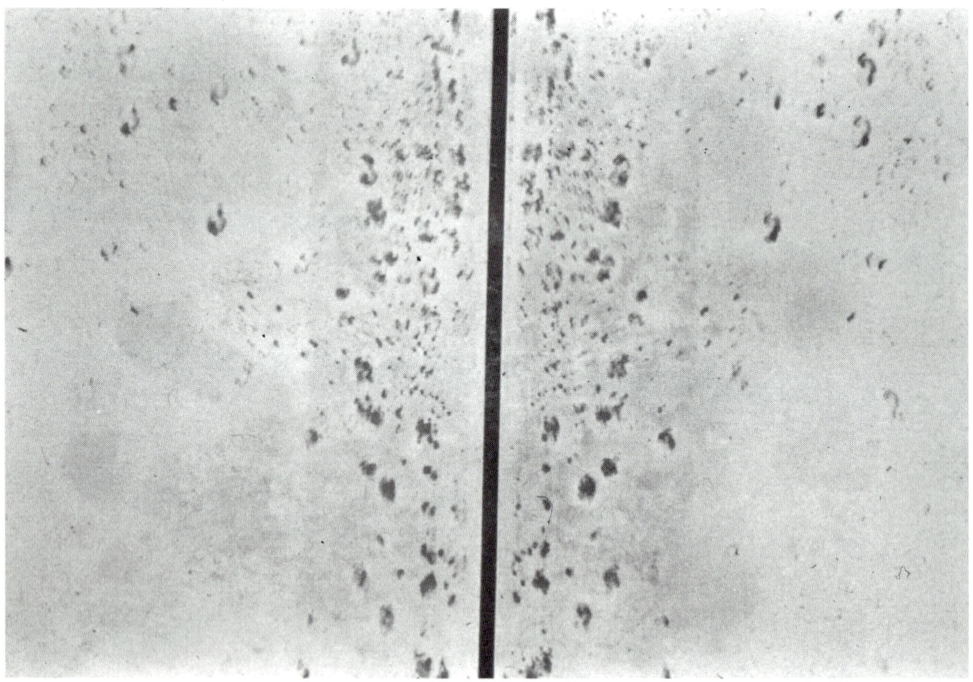

Fig. 37 Attack by fretting corrosion on a brass strip; the attack has arisen in a coil on spots being in contact during lengthy transportation (Metallverken) (x 1/3).

4.11 ENVIRONMENT-INDUCED CRACKING

Environment-induced cracking can arise from the joint action of mechanical tensions and corrosion. If the tensions consist of static tensile stresses, then one refers to stress corrosion cracking; but if the load varies, then the process is called corrosion fatigue. There is no clear borderline between the two processes; both can lead to cracking and fracture. A recent conclusion is that it is the straining caused by the tensions which contributes more to the attack, than the actual tensions.

4.11.1 Stress corrosion cracking

Stress corrosion cracking is caused by static tensile stresses in a material in the presence of a specific corrosive medium. This type of corrosion is best known amongst certain alloys but it has also been found to occur in pure metals under unfavourable conditions.

It is only mechanical tensile stresses above a certain critical level which give rise to stress corrosion cracking. Mechanical compressive stresses are completely harmless. The tensile stress can be:

— a residual stress, which remains from earlier cold deformation, for example bending or deep drawing, or
— an applied stress via direct load.

The corrosive medium which contributes to stress corrosion cracking is to some extent specific to the metal concerned, for example ammonia for copper alloys, chloride solutions for austenitic stainless steel and nitrate solutions for carbon steel. But a number of other substances have been shown to cause stress corrosion cracking under unfavourable conditions. It is often the oxygen content, the pH value and the electrode potential which are decisive.

Within the process of stress corrosion cracking one can distinguish two main stages: *initiation* and *propagation*. Initiation is the process which takes place before any crack becomes visible. Propagation implies the growth of the crack and can result in fracture or in repassivation. In the latter case the process is stopped. The mechanisms vary with the material and the corrosion conditions.

Stress corrosion cracking gives rise to cracks which can be:

— *transgranular*, i.e. passing right through the crystal grains, or
— *intergranular*, localised at the grain boundaries (Fig. 38).

The cracks can cause the piece of material to break. Stress corrosion cracking is characterised by *brittle fracture*. This implies that no contraction takes place at the place of rupture as with a *ductile fracture* (Fig. 39).

Countermeasures against stress corrosion cracking aim at eliminating either the tensile stresses or the corrosive medium or if possible, both. Adjustment of the potential is also effective as may be the addition of inhibitors. A usual countermeasure is *stress-relief annealing*, whereby residual stresses in the construction are reduced to a harmless level. When using stress-relief annealing one chooses the conditions, temperature and time, usually in such a way that the residual stresses are reduced to a satisfactory level without the strength of the material being adversely affected. With brass for example, in many cases heat-treatment at 300°C for 1 hour is suitable; for stainless steel a higher temperature is required (about 500°C).

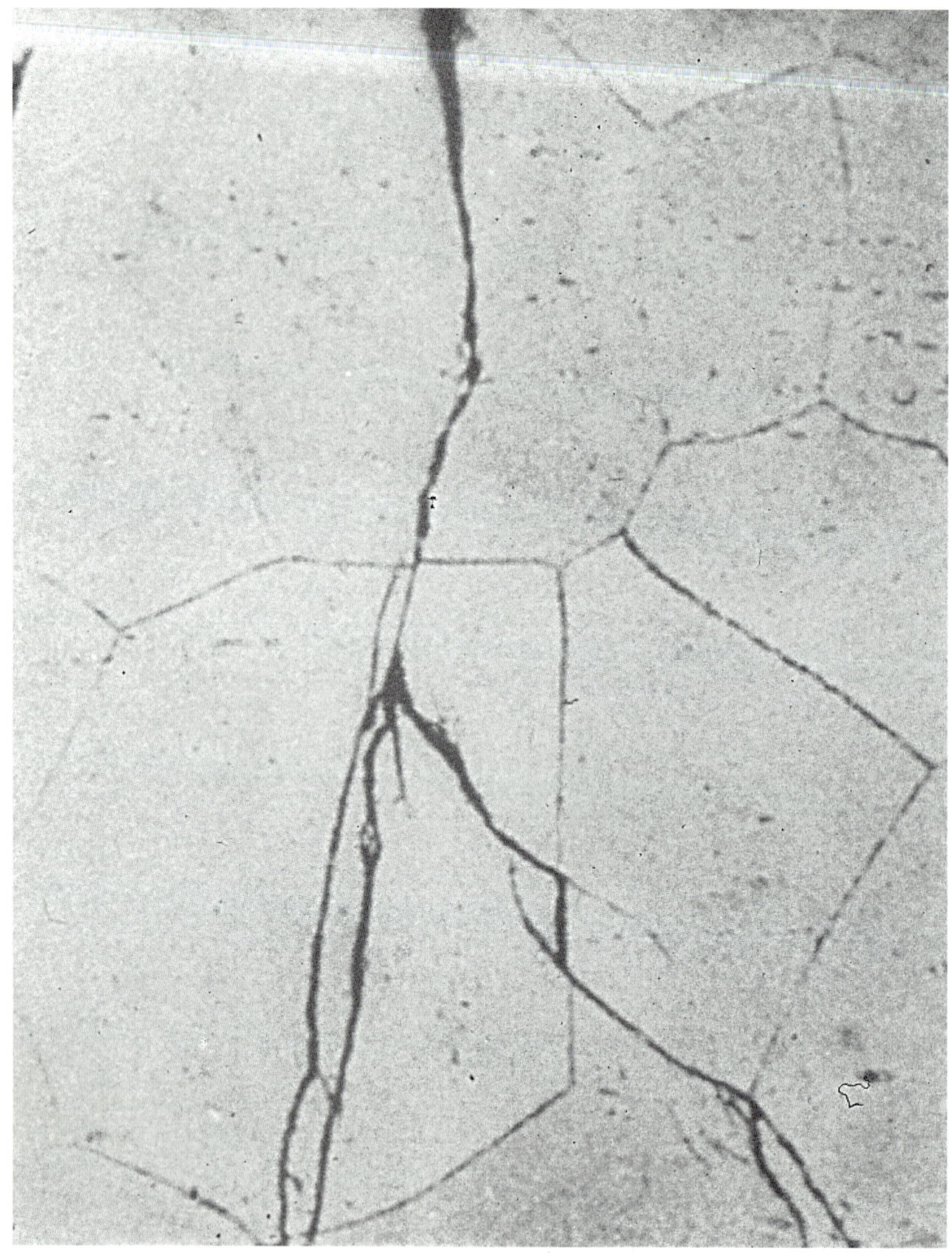

Fig. 38(a) Cross-section showing transgranular cracking (Metallverken) (x500).

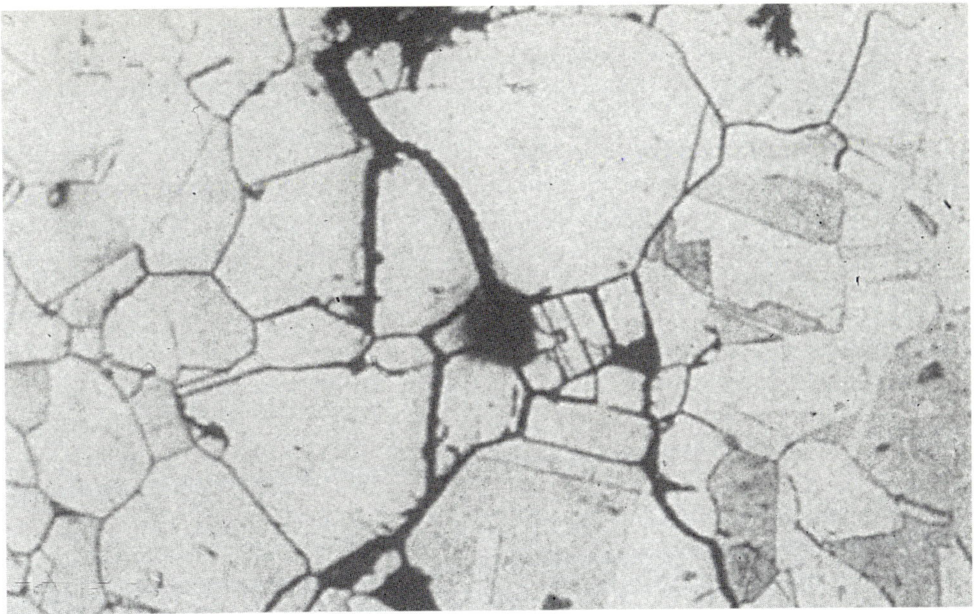

Fig. 38(b) Cross-section showing intergranular cracking (Metallverken) (×500).

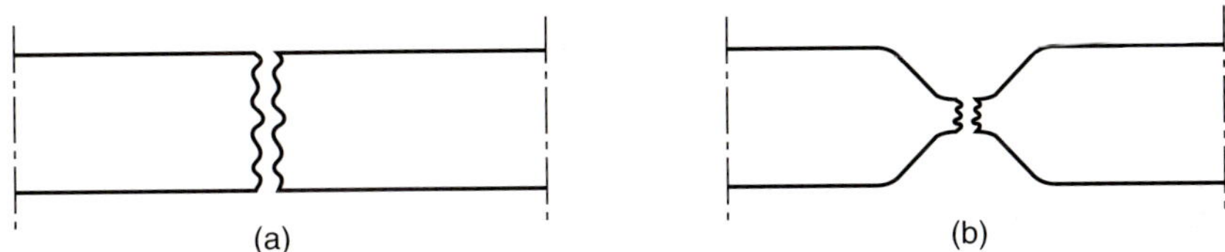

Fig. 39 Brittle (a) and ductile (b) fracture.

Another countermeasure is *shot-peening*, whereby compressive stresses are introduced into the surface zone of the material.

When carrying out *stress corrosion testing* it is a basic rule to simulate both the corrosion environment including the corrosion potential and the stress conditions as close as possible to those in service in order that the corrosion mechanism will be the same. Recommendations on stress corrosion testing are given in the standards ISO 7539/1-7. The following methods can be named:

— *Tests under constant strain.* The test piece which is exposed to the corrosive medium can be, for example, in the form of clamps, U-bends or cups made from sheet material by cold deformation or so-called C-rings made from rods, sections, pipes etc., which before exposure are compressed or widened in a defined manner (Fig. 40).

— *Tests under constant tension load.* In this case the exposed test piece consists of a tensile specimen which is subjected to uniaxial tension with the aid of weights or a calibrated spring. Sometimes several tensile specimens are coupled together in a sequence.

— *Constant strain rate testing.* In a specially built tensile testing machine the test piece, in the form of a tensile specimen, is subjected to straining at a constant rate (10^{-3}–10^{-7} m s^{-1}) while simultaneously exposed to the medium. The straining is continued until rupture. During the process the tension is measured as a function of the elongation (Fig. 41). A test usually takes a couple of days.

— *Determination of fracture mechanics data.* In this case one can use a pre-cracked test piece in which a fine crack has been made mechanically. The test piece, in the presence of the corrosive medium, is put under tension load at right angles to the crack so that the latter tends to widen and grow (Fig. 42). The propagation (V) is measured and recorded as a function of the stress intensity factor (K), an expression of the stress conditions at the tip of the crack and which can be calculated. With the aid of the K-V diagram one can, amongst other things, determine the maximum tension which can be allowed in a construction such that it can be considered to be safe with respect to stress corrosion cracking during a given working life (Fig. 43). Stress corrosion cracking investigations of this type are, however, relatively demanding and are, because of practical limitations, suitable only for materials of low ductility, such as certain stainless steels, aluminium and titanium alloys.

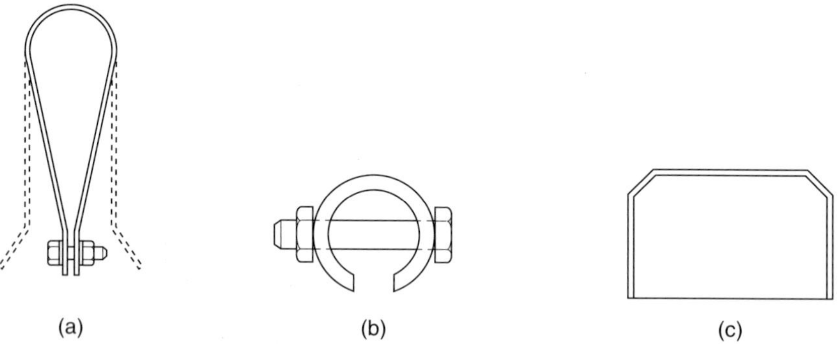

Fig. 40 Test pieces with constant strain for stress corrosion investigations: (a), clamp (the dotted line shows the shape before compression of the clamp). (b), C-ring (tension is induced by tightening the nut). (c), deep-drawn cup with residual stress.

4.11.2 Corrosion fatigue

When a material is subjected to a varying load, changes in the material can occur which result in damage, even if the load is considerably lower than the ultimate strength of the material. Changes in load can in unfavourable conditions lead to crack formation and fracture. This type of damage is called *fatigue*. The rate of attack can

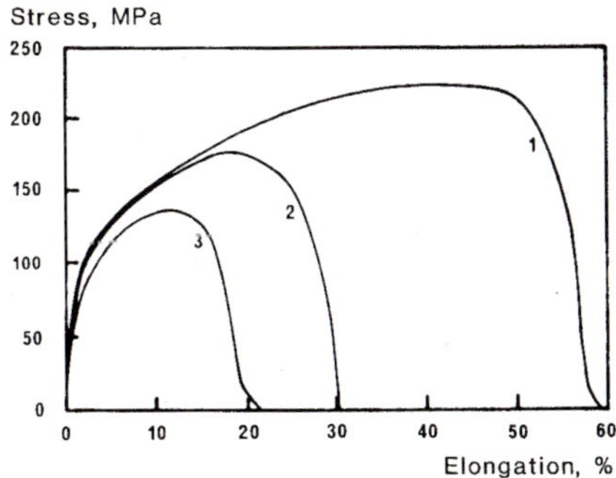

Fig. 41 Stress-elongation curves, obtained on stress corrosion testing of copper using constant strain rate technique. (1) in air (no stress corrosion cracking). (2) in 1 M $NaN0_2$ solution at the free corrosion potential. (3) in 1 M $NaNO_2$ solution at an electrode potential of 100 mV in comparison with a saturated calomel electrode [5].

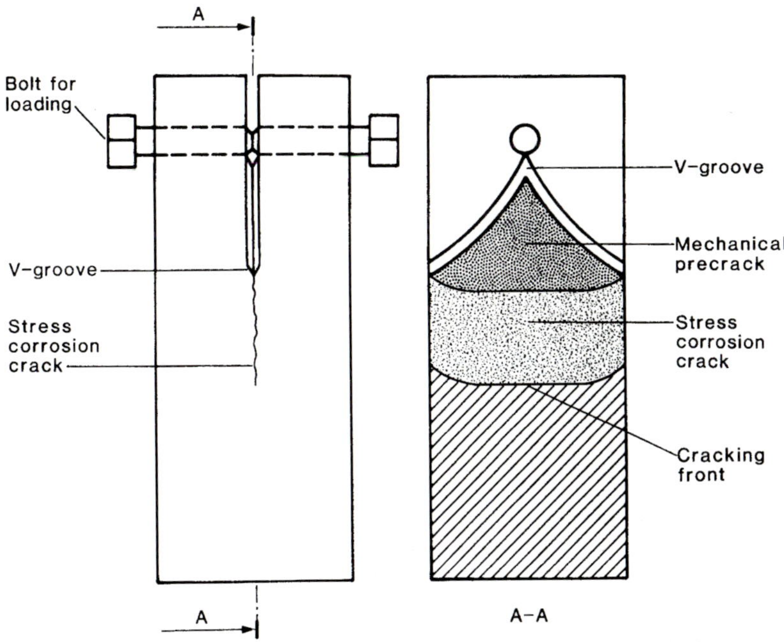

Fig. 42 Test piece used in stress corrosion testing with mechanical fracture technique.

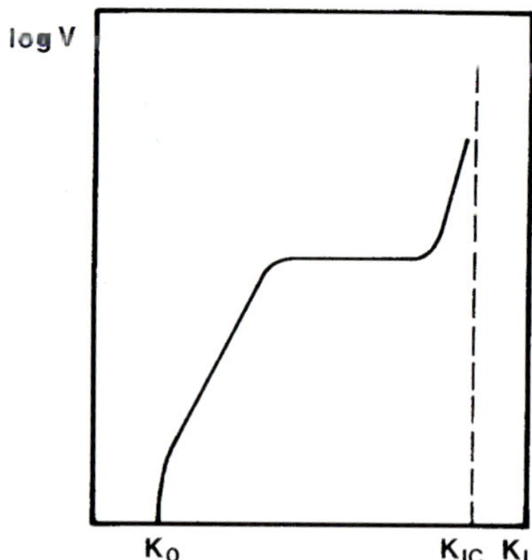

Fig. 43 Schematic K-V diagram obtained on stress corrosion testing with mechanical fracture technique; *V*=propagation rate of the crack, K_I=stress intensity factor, K_{IC}=critical value at which the crack propagation rate is very high, K_O=threshold value under which crack propagation cannot be measured.

be considerably increased if corrosion takes place simultaneously. This phenomenon is called *corrosion fatigue*. In this case no special corrosive medium is necessary as with stress corrosion cracking. The time to fracture is dependent on the number of load changes (*N*) and the magnitude of the load (*S*). Related values of these parameters for fracture can be recorded in a so-called Wöhler curve (Fig. 44). The result is, however, influenced by the frequency of load changes and the form of the load curve (sine, triangular or square wave).

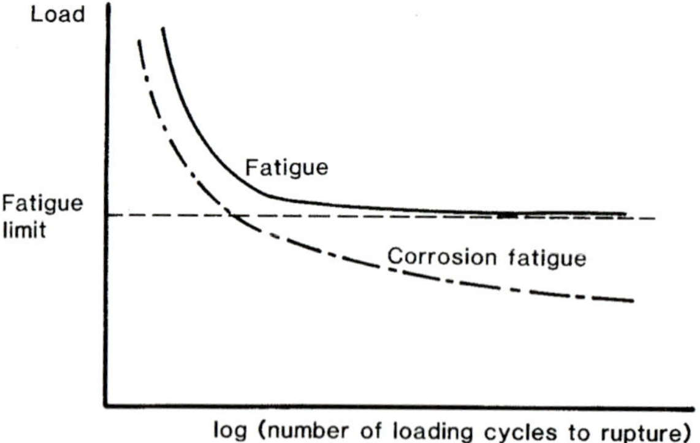

Fig. 44 Wöhler (S/N) curves for fatigue and corrosion fatigue in steel.

With certain metals, e.g. steel, fatigue fracture does not take place irrespective of the number of load changes, providing only pure fatigue occurs, and the load is kept below a given value, which is called the *fatigue limit* (Fig. 44). With *corrosion fatigue*, however, there seems to be no pronounced fatigue limit.

The cracks which appear with fatigue and corrosion fatigue are usually transgranular, straight, unramified and wide. With corrosion fatigue several usually appear together in colonies (Fig. 45), but with 'pure' fatigue there are often only a few cracks.

The fractures resulting from fatigue and corrosion fatigue are, as with stress corrosion cracking, of a brittle type (Fig. 39). On the fractured surface one can often distinguish the spot where the cracking started and a number of curves, which show how the cracking front has spread (Fig. 46).

Damage via fatigue and corrosion fatigue can arise in many different situations. As an example one can mention the cracking in ships' propellers. The load on each propeller blade will vary with the blade's position in relation to the hull. So under running conditions a large number of changes in load occur simultaneously with corrosion by the sea water. In unfavourable conditions, therefore, corrosion fatigue can arise.

Another example is cracking in hot-water systems, e.g. pipes for hot tap water or district heating networks, which have been mounted so they have insufficient elasticity. Temperature variations can give rise to damage-causing variations in load. These can be counteracted with special *expansion joints,* e.g. *horseshoes, loops, glands, bellows* or *bends*, which are included in the system (Fig. 47).

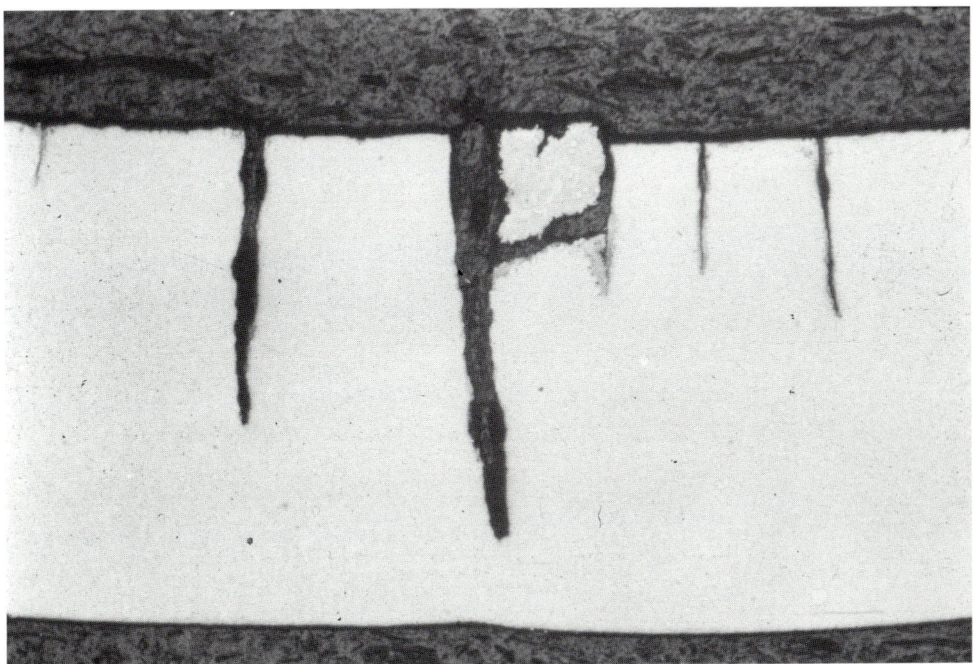

Fig. 45 Cross-section through the wall of a copper pipe with cracks caused by corrosion fatigue (Metallverken) (X50).

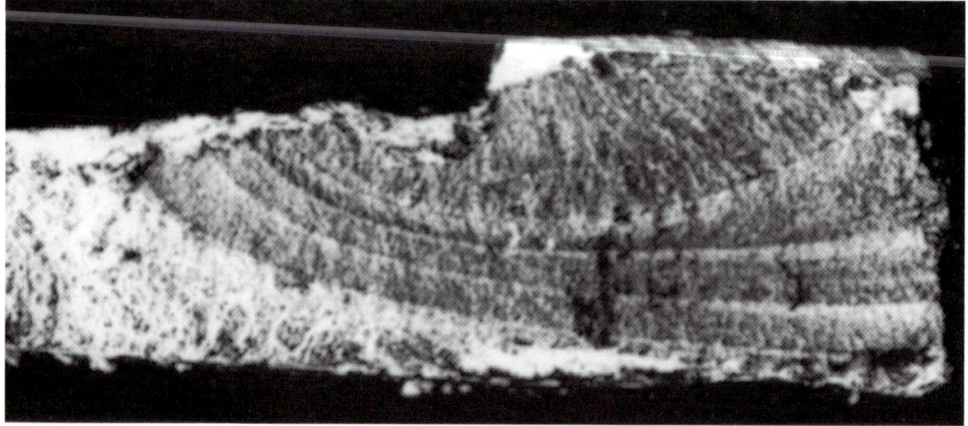

Fig. 46 Appearance of the fracture surface after corrosion fatigue in detail of stainless steel (AISI 316) of paper pulp fibriliser; from the point of initiation the cracking front has moved forward stepwise,which is shown by the characteristic arc-shaped lines (Avesta AB).

Fig. 47 Expansion bend in a district heating line.

4.12 BIMETALLIC CORROSION OR GALVANIC CORROSION

Bimetallic corrosion arises due to the action of a bimetallic cell, i.e. a galvanic cell, where the electrodes consist of different materials. They can be composed of two different metals or of one metal and another electronic conductor, e.g. graphite or magnetite (Fig. 48). Bimetallic corrosion is often called *galvanic corrosion*. This concept has however, a wider meaning and includes also corrosion which results from the action of other galvanic cells, e.g. concentration cells.

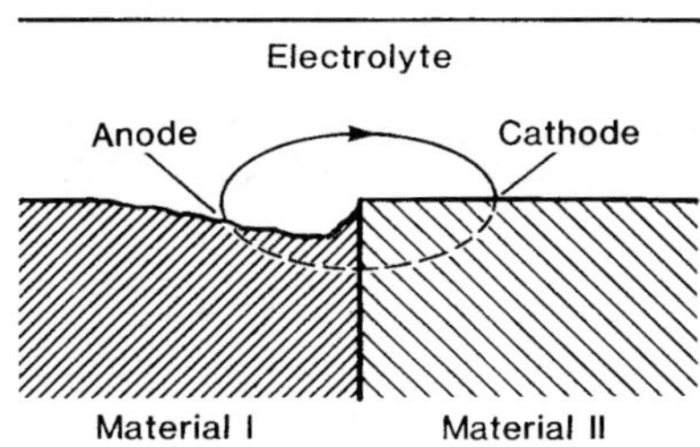

Fig. 48 Corrosion cell in bimetallic corrosion.

As can be seen from Fig. 48, for the corrosion cell to work it is necessary that the contact area between the materials is covered by an electrolyte. In a corrosion cell having two different metal electrodes the more noble metal will be the cathode and the less noble the anode.

The anode reaction consists of a metal oxidation:

$$Me \rightarrow Me^{n+} + ne^- .$$

The cathode reaction consists of a reduction, usually of oxygen dissolved in the electrolyte solution:

$$\tfrac{1}{2}O_2 + H_2O + 2e^- \rightarrow 2OH^- .$$

It is only the less noble material (the anode) which is attacked. Bimetallic corrosion means, in other words, that the corrosion rate of a metal in the presence of an electrolyte is increased via contact with a more noble metal. This is because the reduction reaction takes place more easily on the more noble material than on the less noble one, and the available corrosion current is increased as the cathode area is increased.

Information regarding the nobility of different metals can be obtained from:

— *the electrochemical series*, which applies, (as described in Section 2.6) only to certain conditions, which correspond to the standard electrode potentials (see Table 2))

— *galvanic series*, as determined for different corrosive media, e.g. sea water (see Table 3).

With bimetallic corrosion the relationship between the anode and cathode surface areas is of utmost importance for the degree of attack. As can be seen from Fig. 49 the attack is distributed over a wide area and is therefore insignificant in most cases if

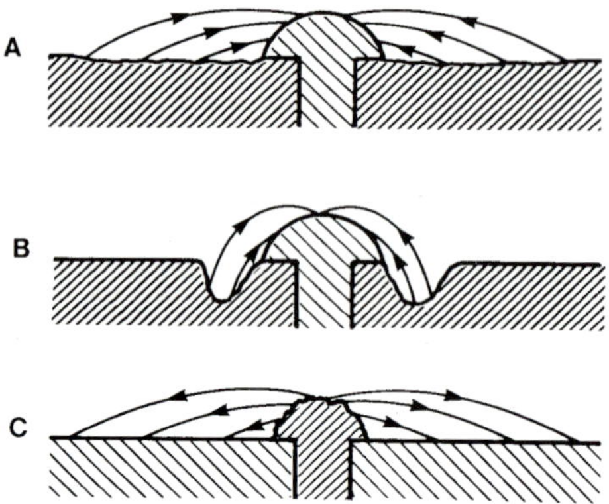

Fig. 49 Bimetallic corrosion in a riveted joint in the presence of an electrolyte, different cases: A, large anode area, small cathode area, good conductivity in the electrolyte. B, large anode area, small cathode area, poor conductivity in the electrolyte. C, small anode area, large cathode area.

the anode area is large in comparison with that of the cathode and if the solution has good electrical conductivity (case A). Should, however, the latter conditions not be fulfilled, a considerable attack can take place in the proximity of the cathode (case B). The risk of serious attack is great, when the anode area is small in comparison with the cathode (case C).

It has previously been mentioned that the presence of an electrolyte is necessary for bimetallic corrosion to take place. If the metal surface is dry, there will be no bimetallic corrosion. On outdoor structures the moisture film which forms on the metal surface can be sufficient for bimetallic corrosion. In cases of a combination of aluminium with copper, steel or stainless steel, it has been claimed that bimetallic corrosion of practical significance takes place primarily in marine atmospheres but is unusual in urban or rural environments. This is probably because marine atmospheres contain a high concentration of chlorides which not only provide good electrical conductivity but can also weaken the protective oxide coating which is normally present on aluminium. In agreement with this there is considerable risk of bimetallic corrosion when the surface is contaminated with, for example, road salt. The risk for bimetallic corrosion in a number of combinations of metals in different atmospheres can be seen in Appendix 1.

To avoid bimetallic corrosion one should observe the following precautions:

— Do not connect metals which are well separated in the electrochemical series or in a more representative galvanic series. This requirement must be fulfilled in marine atmospheres and where the metal surfaces are expected to be permanently exposed to moisture.

— When required insulate, if possible, dissimilar metals from each other so that metallic contact does not occur, for example with age-resistant plastics or rubber (Fig. 50).

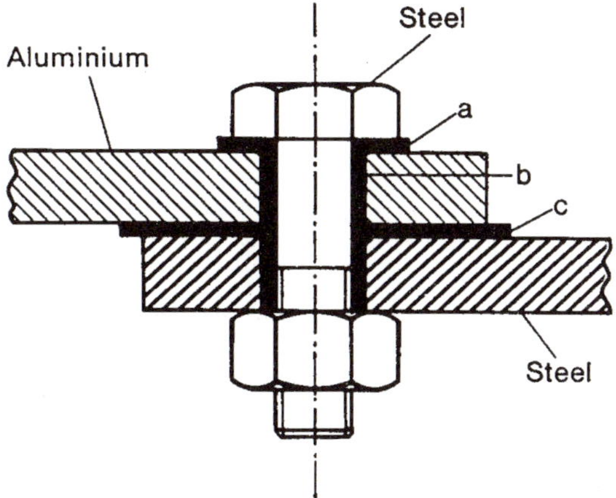

Fig. 50 Protection against bimetallic corrosion in a screw joint of aluminium/steel; a, b, c=insulating material of, for example neoprene rubber.

— Design the construction so that moisture cannot collect and remain at the point of contact.

— Coat the area of contact and its surroundings with corrosion-preventing paint or bitumen. Since the current path will then be longer and the resistance greater, such a measure will lead to the rate of bimetallic corrosion being considerably reduced. Painting should not be limited to the less noble material, since localised corrosion can then take place at the pores of the coating. Painting of only the noble material is sufficient in many cases.

4.13 STRAY CURRENT CORROSION

This type of corrosion — which occurs in underground structures and in water — is caused by stray current from electrical equipment having some current-carrying part in contact with the soil or the water. In general it is only direct current which gives rise to stray current corrosion in iron and steel. Stray current likely to cause damage can originate from, for example, direct current driven tramcars or underground trains, direct current transmission lines or direct current welding equipment. Alternating current driven trains do not generally cause stray current corrosion.

Fig. 51 shows how stray current corrosion can appear in a steel pipeline situated

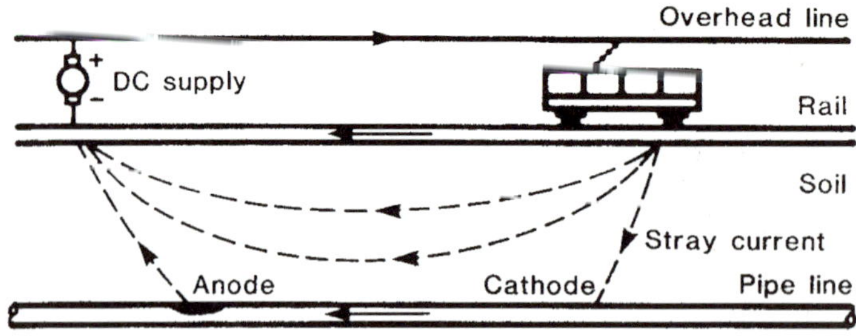

Fig. 51 Stray current corrosion near a direct current railroad.

alongside an electric railroad driven by direct current. Part of the current, the stray current, is not returned to the current source (rectifier station) via the rails as was planned, but instead travels via the soil where the steel pipe is situated. Since the steel pipe is very conductive, part of the stray current is transferred to the pipe. Near the rectifier station the current leaves the pipe and returns to the station. Where the current enters the pipe from the moist soil, the pipe acts as a cathode. Where the current 'leaks out' of the pipe to return to the feeder station an anodic reaction and corrosion of the steel takes place according to:

$$Fe \rightarrow Fe^{2+} + 2e^- .$$

Perforation of the pipe can soon take place and may occur in as little as two years.

In steel and cast iron pipelines the individual lengths of pipe are sometimes insulated from each other with rubber sealing washers. If the potential drop in the pipe is large, then the stray current can 'jump' across the insulated joint, i.e. leave the pipe before the joint and re-enter after it. When the current leaves the pipe it causes localised corrosion.

Countermeasures against stray current corrosion aim at preventing current exiting from the metal and the accompanying anodic attack. This can be achieved by bringing the pipe into electrical (metallic) contact with the negative pole of the interfering current system. This measure is called *electrical drainage*. On *direct electrical drainage* the connection contains no other regulating device than possibly a resistor. This type is used when the stray current always flows in the same direction. If the stray current often changes strength and direction, as for example when it originates from direct current tramcars or direct current welding, then polarised or forced electrical drainage is used. With *polarised electrical drainage* (Fig. 52) a rectifying valve is included, for example, a silicon or germanium diode,which ensures that the current only travels in the desired direction. One can also include into the connection a source of direct current, which gives the protected object a more negative potential compared with the surroundings. This is called *forced electrical drainage* (Fig. 53). Then the current can be continuously regulated with the aid of a reference electrode (copper/copper sulphate) and a potentiometer so that the potential of the protected structure is always held at a suitable level.

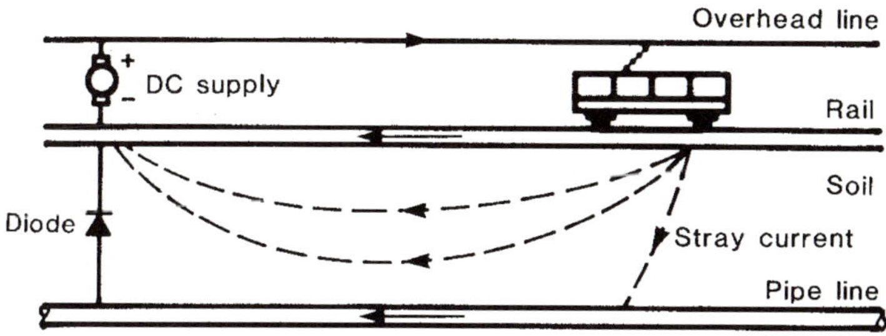

Fig. 52 Polarised electrical drainage.

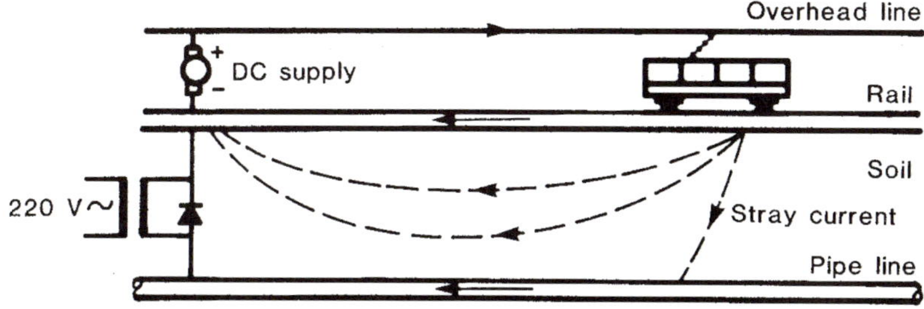

Fig. 53 Forced electrical drainage.

5
Corrosion environments

In ancient times one spoke of the four elements: water, earth, air and fire. To some extent the most important corrosion environments of today: water, soil, atmosphere and dry gases are in agreement with this classification, as problems with dry gases are primarily associated with high temperatures.

5.1 WATER

Many important structures are exposed to water, e.g.:

— mains for warm or cold tap water; pipes, fittings, valves and pumps,
— water cooling systems; pipes, heat exchangers, pumps, etc.,
— central heating systems; pipes, radiators, valves and pumps,
— steam power installations; boilers or steam generators, superheaters, steam turbines, condensers, pipes, valves and pumps,
— ships; hull and propellers,
— harbour installations; often with steel piling, and locks.

Where corrosion is concerned two types of water are usually distinguished: fresh water and sea water. Besides these the purer water found in steam installations should be mentioned, although the corrosion conditions are rather special.

5.1.1 Fresh water

The corrosivity of *fresh water* is dependent on factors such as oxygen concentration, pH value and hardness as well as HCO_3^-, Cl^- and SO_4^{2-} concentrations.

Oxygen concentration

Of decisive importance for the corrosivity of the water is the concentration of free *oxygen*, since this species takes part in the cathode reaction:

$$\tfrac{1}{2}O_2+H_2O+2e^- \rightarrow 2OH^- \ .$$

If oxygen is missing there is usually no oxidising agent. In consequence the cathode reaction and therefore the whole corrosion process is eliminated. The solubility of oxygen in water which is in contact with air falls off with temperature as can be seen in Table 7. In closed hot water systems the concentration of oxygen will fall as a result of

Table 7 Solubility of oxygen in water in contact with air at a pressure of 1 atm and different temperatures

Temperature (°C)	Solubility (mg l^{-1})
0	14.6
20	9.1
40	6.4
60	4.7
80	2.8
100	0.0

the reduced solubility at the higher temperature and reaction with the large area of ferrous metals (radiators, pipework etc.) in the circuit and will usually stabilise at a low level, for example at a few tenths of a milligram per litre. This will be the case providing that the amount of make-up water is not unusually great and oxygen is not available through, for example, oxygen-permeating walls of plastic pipes, a badly constructed expansion vessel or a defective circulation pump. By dosing an *oxygen scavenger*, e.g. sulphite or hydrazine, the level of oxygen can be reduced even further (see 5.1.3). In closed central heating systems radiators made of steel can be used with brass fittings and steel pipes, and often with copper pipes, without significant corrosion arising. But with oxygen-rich water, such as distribution water, the corrosion rate of steel pipes is often considerable. Furthermore, mixed metal installations of, for example, steel and copper may increase the risk of corrosion of the steel. The influence of oxygen on corrosion can be seen on components which are only partly submerged in water, where as a rule the strongest attack takes place immediately below the surface of the water (Fig. 54), where the supply of oxygen is greatest. Localised corrosion of this kind is called *water-line corrosion*.

pH value

In water having a *pH value* less than about 4, i.e. in acidic solution, carbon steel will corrode at a considerable rate, even in the absence of oxygen. Under such conditions H^+ ions can act as the oxidising agent according to the formula:

$$H^+ + e^- \rightarrow \tfrac{1}{2}H_2 \ .$$

However, the pH value of distribution water is usually about 8. If it is considerably lower, there is often a risk of corrosion in the pipes, since corrosion products such as oxides, hydroxides or hydroxide salts, which can provide a protective coating at pH values above about 7, cannot form at low pH values.

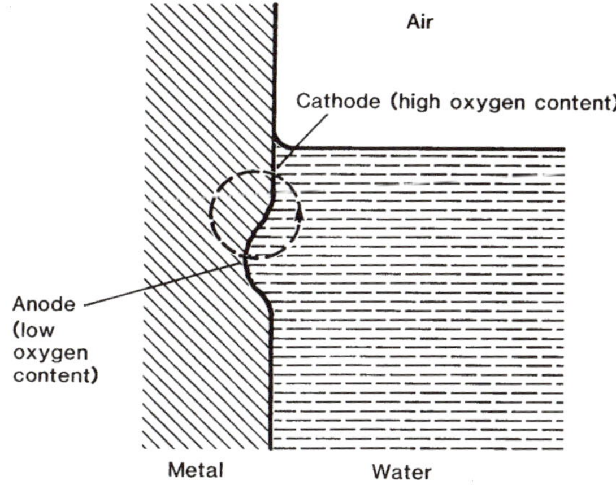

Fig. 54 Water-line corrosion.

Precipitation of calcium carbonate

The risk of corrosion is also dependent on the possibilities for precipitation of calcium carbonate according to the formula:

$$Ca^{2+} + HCO_3^- + OH^- \rightarrow CaCO_3 + H_2O \ .$$

Calcium carbonate, together with corrosion products that may be formed, can produce a protective scale on the metal surface. At pH values higher than the so-called *saturation pH* (pH_s) the water can deposit calcium carbonate. Conversely, at lower pH values it does not possess this characteristic and will dissolve any existing calcium carbonate precipitate. The saturation pH at a given temperature can be calculated if the concentrations of Ca^{2+} and HCO_3^- as well as the total salt content are known. In order to determine whether or not the pH value of the water is above or below the saturation pH one can calculate a saturation index, e.g. *Langlier's index* (*L*):

$$L = pH - pH_s.$$

The calculation of Langlier's index can be carried out with the aid of the nomogram in Appendix 2.

If *L* has a positive value, the water is scale-forming, if on the contrary the value is negative, the water is not scale-forming and as a consequence is often corrosive.

Acidification

Acidification of the environment by acid rain leads first to the consumption of buffering substances in the water, mostly HCO_3^-, and in extreme cases to a reduction in pH. Both of these changes adversely affect the requirements for the formation of a protective coating on the metal surface and thereby increase the corrosivity of the water. The effect is most noticeable in water of low buffer capacity, as found, for example, in Sweden in the county of Småland and along the west coast (Fig. 55).

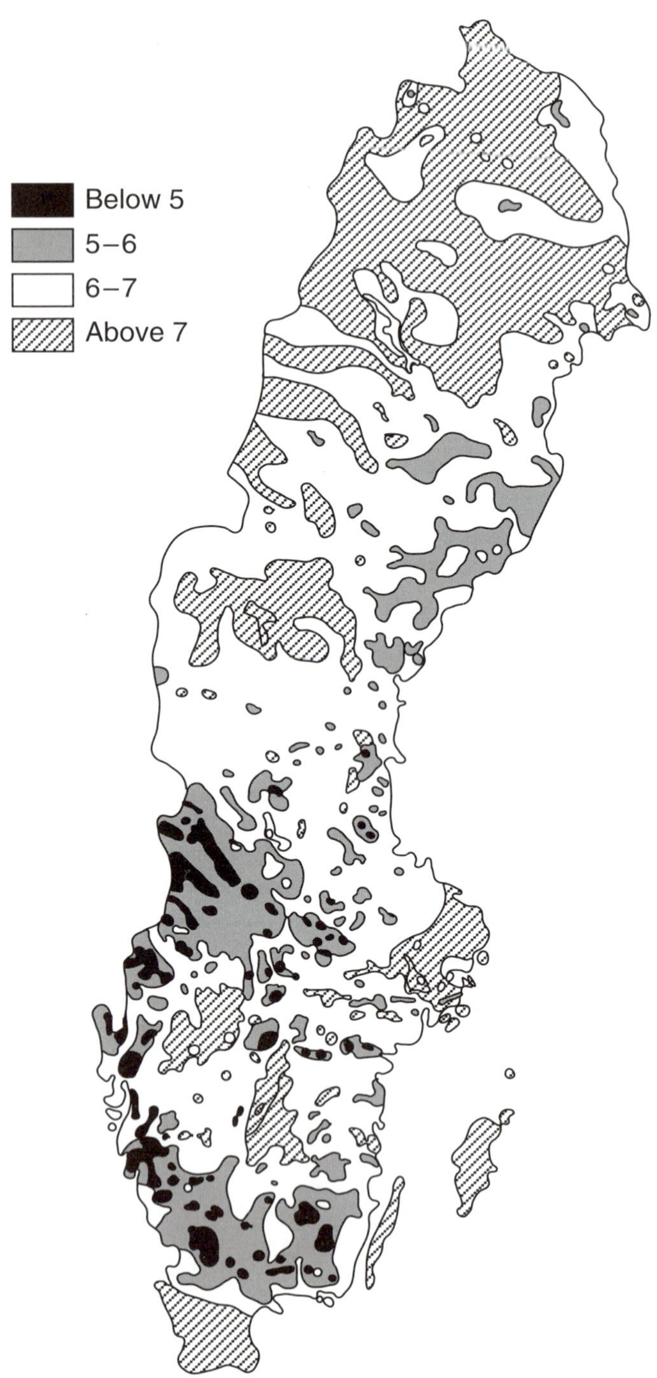

Fig. 55 pH values in Swedish lakes winter and spring [7].

Alkalisation and carbonation

At certain waterworks *alkalisation* (increase in pH) and/or *carbonation* (increase in the concentration of HCO_3^-) of the water is carried out to reduce its corrosivity. According to Swedish building norms tap water should, from the corrosion viewpoint, have a pH value of between 7 and 9. With regard to the HCO_3^- concentration it should be at least 70 mg l^{-1}. In waterworks alkalisation is usually made by the addition of sodium hydroxide (NaOH) or lime $Ca(OH)_2$) or by passing the water through a bed of burnt dolomite ($CaMg(CO_3)_2$), the so-called magno mass. Carbonation can be achieved by the addition of lime and carbon dioxide (CO_2). Well water can be made alkaline with the aid of magno mass.

Flow velocity

The flow velocity of water influences corrosion. In stationary water conditions pitting can arise (see 4.2). This situation is associated partly with inadequate renewal of water across the metal surface and partly with an insufficient supply of the substances required for the production and maintenance of a protective coating. But with too high a water velocity, on the other hand, there is risk of erosion corrosion (see 4.8).

Temperature

An increase in temperature usually brings about an increase in the rate of corrosion. A 10°C rise in temperature usually doubles the rate of chemical reactions and also increases the rate of diffusion controlled processes. But in many cases the corrosion rate reaches a maximum at about 80°C. On further temperature increase it often goes down because the solubility of oxygen decreases (Table 7).

5.1.2 Sea water

Clean *sea water* in the large oceans has only small variations in composition and corrosivity. Its pH value does not deviate much from 8.1 and its salt concentration is about 3.5 weight per cent, mostly as NaCl. But in harbours and other places near land sea water can have a different composition. This can be because of the inflow of river water or the release of polluted sewage. In the Baltic, for example, the concentration of NaCl decreases as the distance from the Atlantic increases (Fig. 56). Harbour water often contains sulphur compounds, which increase the corrosivity of the water considerably. In corrosion testing it has been difficult to produce a synthetic sea water with the same corrosivity as natural sea water. An important reason is that natural sea water contains microorganisms which are missing in synthetic sea water and which can influence corrosion.

Metallic objects exposed to sea water can acquire *growths* of sea organisms, e.g. barnacles, mussels and, in the presence of daylight, also algae (Fig. 57). These growths, generally called *fouling*, can cause deposit corrosion (see 4.4). Other forms of damage can also result, e.g. the blocking of pipes and the increase in frictional drag on ships. On the other hand such growths can, under certain conditions, give rise

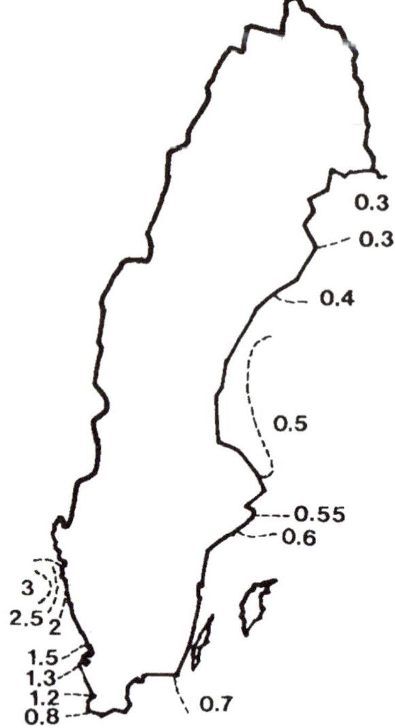

Fig. 56 The concentration of salt (%) in sea water along the Swedish coast.

to corrosion protection, e.g. on steel. In water-pipes growths can be prevented by *chlorination* using, for example, sodium hypochlorite solution or chlorine gas, which is added at the water inlet. Growths on ship's hulls can be hindered by painting with *anti-fouling paint*. This gives off species which are poisonous to sea organisms, e.g. copper ions or tin compounds. On a copper surface the tendency for fouling is negligible. The copper which dissolves on corrosion acts as an anti-fouling agent.

There are powerful movements in the sea which contribute to a concentration equalisation. But in spite of this there are distinct differences in corrosion conditions:

— Because of the limited light penetration the type and extent of the growths change with depth and these changes will affect corrosion.
— A sea current can have a different flow velocity and temperature than its surroundings, and therefore produce corrosion rates differing from those in the surrounding mass of water.
— In certain parts of the sea and at great depths a so-called pycnocline exists which divides and hinders convection between the water at the bottom and that above. In, for example, Gullmarsfjorden on the west coast of Sweden the pycnocline is at a depth of about 30 m. The water at the bottom consists of water from the North Sea, whilst the zone above has a lower salt content as a consequence of dilution by surface water from watercourses which open into the fjord.

Fig. 57 Testing frame with fouling; the frame has been exposed for about three years to sea water on a testing raft at Bohus Malmön (Skagerack).

5.1.3 Water in steam generation plants

There are special requirements for water chemistry in steam generation plants. The main components can be seen in the simplified diagram in Fig. 58. Steam is produced

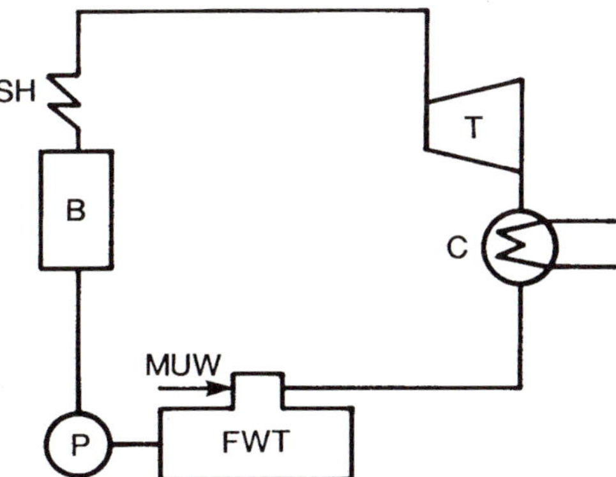

Fig. 58 A schematic diagram of a steam power plant. B, boiler; SH, superheater; T, steam turbine; C, condenser; MUW, make-up water; FWT, feed water tank; P, pump.

in a boiler or steam generator (B). After raising the temperature in the superheater (SH) part of the energy content is exploited in a steam turbine (T) or a steam engine. After this the steam passes into a heat exchanger, condenser (C), where condensation takes place via heat transfer to cooling water. After any water losses have been replaced by addition of make-up water (MUW) in a feed water tank (FWT), the condensate is returned to the boiler/generator.

Water chemistry in steam generation plants must be adjusted so that deposits and corrosion are avoided in the different units of the plant. This requires a careful control and treatment of water that is supplied to the boiler/generator, with regard particularly to oxygen, Ca^{2+}, Mg^{2+}, Cu^{2+}, HCO_3^-, SO_4^{2-} , Cl^- and silicates a well as the pH value. The following types of treatment are used: softening, deaeration, pH control and addition of inhibitor.

Softening

Softening aims at reducing the concentration of Ca^{2+} and Mg^{2+}, which can give rise to deposits. Several processes can be used for softening the make-up water, e.g.:

— *thermal softening*, whereby $CaCO_3$ and $MgCO_3$ are precipitated according to the following reaction

$$Ca^{2+} + 2HCO_3^- \rightarrow CaCO_3 + CO_2 + H_2O \quad ,$$

— *ion exchange in sodium-saturated ion exchange filter*, whereby Ca^{2+} and Mg^{2+} are exchanged for Na^+, and

— *total demineralisation*, whereby cations as well as anions are removed with the aid of ion exchangers.

Deaeration

Deaeration aims to reduce the corrosivity primarily at removal of oxygen which is dissolved in the water. Deaeration is required not only for the make-up water, but also for the condensate, since some oxygen leakage usually takes place in the steam generation plant. One can distinguish:

— *thermal deaeration* whereby the water is treated with steam in a separate unit; this method is based on the fact that the solubility of oxygen decreases with increased temperature, and

— *chemical deaeration* in which the oxygen dissolved in the water is removed by reaction with sodium sulphite (Na_2SO_3) or hydrazine (N_2H_4) according to the following formulae:

$$2Na_2SO_3 + O_2 \rightarrow 2Na_2SO_4$$

and

$$N_2H_4 + O_2 \rightarrow 2H_2O + N_2 \quad .$$

pH regulation

pH regulation is aimed at neutralising the carbon dioxide in the units subsequent to the boiler in order to minimise corrosion. A special requirement is that the pH regulating chemicals follow the steam in suitable quantities into the different units. In other words they have to be in the vapour state at the temperature concerned. Ammonia, morpholine and cyclohexylamine are used for pH regulation. Hydrazine can also have a neutralising effect, since at high temperatures it breaks down to produce ammonia according to:

$$3N_2H_4 \rightarrow N_2 + 4NH_3 \;.$$

When water treatment is carried out using only volatile chemicals one speaks of AVT chemistry (All Volatile Treatment).

Addition of inhibitor

Addition of inhibitor is done to counteract corrosion. For this purpose film-forming amines are used, e.g. octadecylamine. For corrosion protection in the boiler phosphate additives are used. Besides its inhibitive function phosphate has the added ability of precipitating the remaining hardness as a sludge which is not deposited in the turbines. Na_3PO_4 is also added in order to raise the pH value.

To maintain the concentrations of dissolved salts and sludge in the boiler water at an acceptable level the so-called *blow down* treatment is carried out on a regular basis, i.e. boiler water is drawn off.

This description of water in steam power plants is limited to the main principles. In actual practice a large number of variations in terms of equipment, operating conditions and cleaning methods are to be found.

5.2 SOIL

Many constructions vital to the community are placed underground and thus exposed to the corrosive action of soil, e.g:

— *water mains and sewers*, are pipes constructed from steel or cast iron; as a result of leakage caused by corrosion a large proportion of the tap water is lost; in some catastrophic instances whole communities have been left without a supply of water due to corrosion damage,

— *steel pipelines for natural gas*; perforation cannot, because of the risk of fire and explosion, be tolerated in such pipelines; therefore the security demands made on corrosion protection are great,

— *steel tanks for fuel oil and gasoline*; these must be corrosion protected because damage to the environment as a result of leakage must be avoided,

— *lead-sheathed telecommunication cables*; if the lead sheathing is perforated then moisture in the soil can enter and cause a short-circuit thus disturbing tele- and computer communication,

— *the steel foundations and anchoring stays to power pylons*; besides the direct costs of repairing corrosion-damaged pylon foundations and stays, a high-tension pylon breakdown caused by corrosion can lead to a drastic disturbance in the supply of electricity to the country,
— *road culverts* made of steel sheet at the junctions of roads or railways and waterways; perforation can result in erosion damage in the road-bed,
— *steel piling* for the foundations of, for example, bridges and buildings in soil susceptible to settling.

5.2.1 The nature of soil

The loose deposits which exist between the ground surface and bedrock consist of soil. The different kinds of soil have been formed from the ice-age up to the present time. Certain kinds of soil have been formed by the disintegration of rock, others by the decomposition of plants and animals. In some cases the matter has been transported by the wind or by watercourses. Soil consists of three components:

— *soil particles*, which can be inorganic in character (e.g. clay, sand and gravel) or organic (e.g. peat and mud); the inorganic types of soil are divided according to particle size; in clay for example they are less than 0.002mm, in silt 0.002-0.06 mm and in sand 0.06-2.0 mm,
— *soil moisture*, which contains different dissolved species, e.g. oxygen, H^+, Cl^-, SO_4^{2-}, and HCO_3^-, and
— *a gas phase*, consisting mostly of nitrogen and oxygen which permeate down via non-waterfilled pores and cracks, as well as carbon dioxide formed by the breakdown of organic substances in the soil.

Soil is seldom homogeneous but varies in both the horizontal and vertical directions. The heterogeneity is increased by digging. In addition, the characteristics change with the seasons as a result of rainfall, thawing of snow, drying etc. On the whole this is the character of the environment to which underground structures are exposed.

5.2.2 Corrosion mechanisms

Even corrosion in soil takes place via electrochemical cells. Soil moisture constitutes the electrolyte and the following electrode reactions are common as in other cases of aqueous corrosion, e.g. in steel:

anode reaction:	$Me \rightarrow Me^{n+} + ne^-$
cathode reaction:	$\frac{1}{2}O_2 + H_2O + 2e^- \rightarrow 2OH^-$
reaction in the soil:	$Me^{n+} + nOH^- \rightarrow Me(OH)_n$.

When the cathode and anode are closely situated and the pH value of the soil is greater than about 5, the corrosion products can form a coating which will provide some protection to the steel surface. The corrosion effect is therefore evenly distributed and the rate decreases with time. Under certain conditions, however, the anode and cathode can be more widely separated and in extreme cases on a pipeline or a cable by as much as one or two kilometres. The metal ions formed at the anode will migrate with the current towards the cathode and the OH^- ions formed at the cathode will migrate towards the anode. The corrosion products are then precipitated somewhere between the anode and cathode. They therefore do not provide any protective coating to the anode. As a result pitting can occur at the anode, with the surfaces showing shiny metal. Since no protective coating forms at the anode, the corrosion rate does not decrease with time, but on the contrary, can increase due to enrichment of ions and increase of conductivity via the action of the corrosion cell. If the cathode area is much larger than that of the anode, then the anodic current density will be high resulting in a high rate of pitting. Localised corrosion can lead to perforation and major damage in, for example, oil tanks or pipes whereas single pits are of less importance for load-bearing structures such as steel piling.

5.2.3 Factors influencing corrosion in soil

Corrosion in soil is influenced primarily by the following factors: the presence of soil moisture, the supply of oxygen, the redox potential, the pH value and the resistivity of the soil and also microbial activity.

Water

The presence of water is a prerequisite for the functioning of corrosion cells, since water usually makes up the major part of the electrolyte. Hence the position of the structure in relation to the *ground-water table* is of major importance for corrosion. In Sweden, for example, the ground-water table is normally located 1–3 metres below ground level — depending on ground conditions and the season of the year — but in well-drained soil such as sand or gravel it can be considerably deeper. Even in the soil above the ground-water table there is water held by capillaries and pores; the finer the soil particles and pore size the more water is held. Water is also supplied via rain, thawing of snow etc.

Oxygen supply

Oxygen takes part in the cathode reaction and a supply of oxygen is therefore a prerequisite for corrosion in soil. The supply of oxygen is comparatively great above the ground-water table but is considerably less below it. It also changes with the type of soil; it is, for example, high in sand but low in clay. Furthermore, the supply of oxygen is considerably greater through fine-grained soil which has been moved, for example, by excavation, than in the same soil in the undisturbed, natural state. If a long structure such as a pipeline crosses two or more different types of soil, e.g. sand and clay, having different characteristics regarding oxygen permeation, then a concentration cell, i.e. an aeration cell, can be formed (Fig. 59). In such a cell the anode is located at a part where the supply of oxygen is low and here localised corrosion will occur. Corrosion cells can, for a similar reason, arise when the structure is surrounded by mixed soil containing, for example, lumps of clay. Pitting occurs under the lumps, where they are in contact with the metal (Fig. 60). The

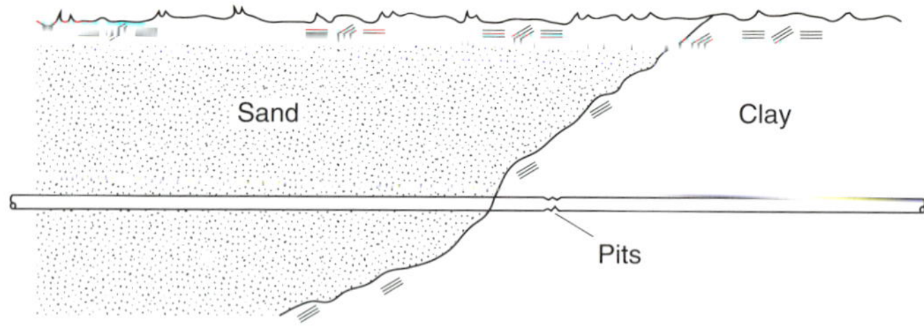

Fig. 59 Corrosion attack on a steel pipe at the transition between clay and sand.

corrosion risk is generally greatest in a zone just above the groundwater table, where the soil pores are only partly filled with water. There sufficient electrolyte is present as well as supply of oxygen from the air for the corrosion cells to operate (Fig. 61).

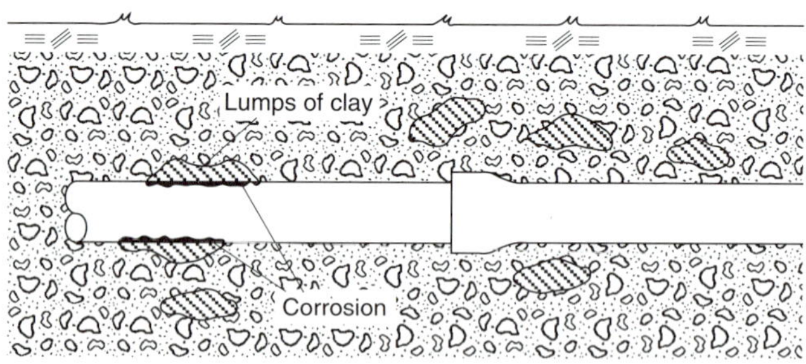

Fig. 60 Corrosion attack under the clay-covered parts of a steel pipe in sand containing lumps of clay.

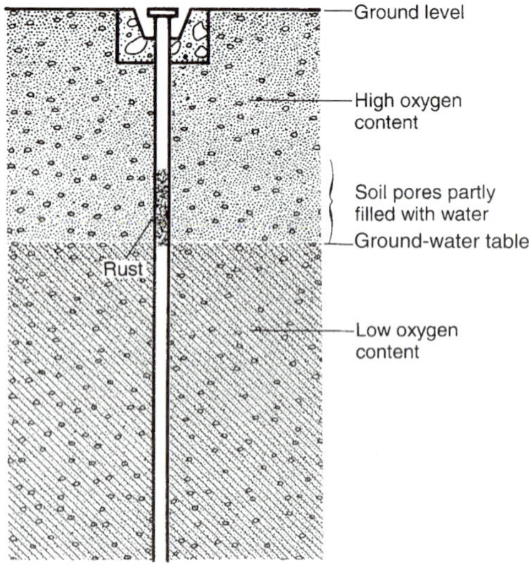

Fig. 61 Corrosion attack on a steel pole in soil just above the ground-water table; varying lengths of the pole will be exposed to corrosion as the ground-water table changes with the season of the year.

Redox potential
The oxygen concentration of the soil moisture generally will determine its redox potential. Thus, the higher the oxygen content the higher the redox potential in accordance with the Nernst equation (see 2.3).

pH value
The pH value of soil moisture affects the solubility of the corrosion products. At pH values below about 4, which can occur, for example, in peat or muddy soils, a protective coating of rust cannot form on steel (see the potential-pH diagram for Fe-H_2O in Fig. 110), and in consequence the corrosion rate can be relatively high. At normal pH values (pH 5–8) the corrosion rate of steel is, however, determined by other factors. The composition of soil moisture can change as a result of acid rain.

Several buffer systems are present in the soil (19), consuming hydrogen ions that have been added:

Buffering system	pH range	Examples of buffering reactions
Carbonate	8.0 - 6.2	$CaCO_3 + H_2CO_3 = Ca^{2+} + 2HCO_3^-$ $CaCO_3 + 2H^+ = Ca^{2+} + H_2O + CO_2$
Organic matter	8.0 - 3.0	$\text{-COOCa}_{0.5} + H^+ = \text{-COOH} + 0.5Ca^{2+}$
Clay material	8.0 - 5.5	$\text{-AlOCa}_{0.5} + H^+ = \text{-AlOH} + 0.5Ca^{2+}$
Silicate	6.2 - 5.0	$CaAl_2Si_2O_8 + 2H^+ + H_2O = Ca^{2+} + Al_2Si_2O_5(OH)_4$
Cation exchange	5.0 - 4.2	$\text{Clay-Ca} + 2H^+ = \text{Clay-H}_2 + Ca^{2+}$
Aluminium	4.2 - 3.0	$AlOOH + 3H^+ = Al^{3+} + 2H_2O$
Iron	3.8	$FeOOH + 3H^+ = Fe^{3+} + 2H_2O$

Each buffer system is primarily effective in a specific pH range. Not until the buffer has been used up will the further addition of hydrogen ions lead to a decrease of the pH value.

Resistivity
The resistivity of soil moisture is determined by the concentrations of the different ions and their mobilities. High levels of salt concentration occur in areas which have previously been sea beds or where large quantities of fertiliser, road salt etc. have been applied. Resistivity influences the current in the corrosion cell, but only where the distance between the anode and the cathode is so great that the IR drop in the cell is of significance. Where the prerequisites for such a cell exist, the risk of localised corrosion (at the anode) is great if the soil resistivity is below 1000 ohm cm, but little if it is greater than 5000 ohm cm.

Microorganisms
In the presence of certain bacteria corrosion can also occur deep in soil having a low

oxygen concentration, i.e. under anaerobic conditions. Certain sulphate-reducing bacteria, e.g. *Desulfovibrio desulfuricans*, can catalyse the reduction of SO_4^{2-} ions, present in soil, to S^{2-} whereas in the absence of bacteria this is a very slow process. This reduction makes possible a corresponding oxidation of steel in such environments. Leaving aside various reaction steps one can represent the net reactions with the following formulae:

anode reaction:	$4Fe \rightarrow 4Fe^{2+} + 8e^-$
cathode reaction:	$SO_4^{2-} + 8H^+ + 8e^- \rightarrow S^{2-} + H_2O$
reaction in soil:	$Fe^{2+} + S^{2-} \rightarrow FeS$
overall rection:	$4Fe + SO_4^{2-} + 8H^+ \rightarrow 3Fe^{2+} + FeS + 4H_2O$.

As can be seen from the overall reaction the corrosion products include FeS and this is a characteristic of this type of corrosion. The presence of FeS can be easily demonstrated by the smell of hydrogen sulphide (H_2S) on the addition of a few drops of dilute hydrochloric acid (HCI).

Corrosion made possible, or accelerated, by microorganisms is called *microbial corrosion* or *microbially induced corrosion* (MIC). Various mechanisms have been suggested. Many microorganisms require the following conditions for their growth:

— a pH value of 5–9,
— the presence of an organic substance,
— a temperature below 40°C, and
— in many cases — a low redox potential.

These conditions prevail in certain soils and also in other environments, e.g. sea water (pH 8.1).

Corrosion in underground metal structures is sometimes also caused by stray current from electrical installations (see 4.13) or by contact with a more noble metal (see 4.12).

5.2.4 Estimation of corrosion risk in soil

Underground metal structures are usually expected to have a long working life, often 50–100 years. But before such a structure is put on site the risk of corrosion and the need for corrosion protection measures should be estimated. The corrosivity of the soil can be estimated by measurement of the parameters discussed in section 5.2.3. A reliable estimate, however, requires a great deal of experience.

Sometimes it is of interest to estimate the corrosion in a metal structure already buried in the soil. A first measure is to make an estimate of the corrosivity of the soil. In addition one can obtain information about on-going corrosion by measurement of the electrode potential of the structure as well as of possible corrosion currents in the surrounding soil. Electrode potential measurements can reveal if there are concentration cells, bimetallic cells or stray currents. These measurements are carried out with the aid of one or more reference electrodes, usually of the copper/copper sulphate type, which are placed on the ground above the structure. The carrying-out of these measurements and the interpretation of the results demands a great deal of experience.

The metals lead and copper have relatively good resistance to corrosion in soil, whereas steel and aluminium are less resistant.

5.2.5 Measures against corrosion in soil

There are several countermeasures to corrosion in soil, e.g.:

- cathodic protection with sacrificial anodes or impressed current (see 6.1),
- electrical drainage so that stray current corrosion is avoided (see 4.13),
- organic coatings, e.g. bitumen, plastic or tape wrapping (see 6.6 and 6.7),
- inorganic coatings, e.g. zinc, zinc-aluminium alloy, or concrete,
- rust allowance, which means a degree of over-dimensioning the structure so as to take the expected corrosion loss into account,
- embedment in concrete, and
- filling the surroundings with sand or gravel; this last procedure is adopted in an attempt to avoid corrosion cells arising from differences in oxygen supply; on the other hand clay, slag, peat or other materials rich in acid, sulphate, chloride or organic matter should be avoided; the effect of filling the surroundings with sand or gravel should not, however, be overestimated.

5.3 ATMOSPHERE

By atmospheric corrosion is usually meant the corrosion which takes place on exposure to the earth's atmosphere at normal temperatures. Many metal structures are exposed to an outdoor atmosphere and thereby also to corrosion. Here may be mentioned metal components in buildings, such as roofs, façades, fastening devices, door and window frames as well as guttering and down-pipes. Other examples are bridges, pylons and vehicles of various kinds. Corrosion rates are usually less indoors but can be significant, e.g. where condensation occurs. In recent years a special type of indoor corrosion has become of great importance, namely corrosion of the components of electronic equipment. Here, corrosion can cause serious breakdown even after relatively little attack.

5.3.1 The effect of moisture

Atmospheric corrosion of metals is an electrochemical process, which takes place in corrosion cells having anodes and cathodes. In order for the corrosion cells to function the presence of an electrolyte is required, which means that the surface must be covered with a moisture film of sufficient thickness, i.e., it must be 'wet'.

The amount of water on rain-protected metal surfaces is to a large extent dependent on the relative humidity of the air, i.e. the ratio of the actual water vapour pressure to the saturation pressure. Below a certain value of relative humidity, the *critical humidity*, the moisture film is so thin that the corrosion rate is negligible in most instances. Above the critical humidity corrosion increases noticeably with increasing relative humidity. The critical humidity depends on the metal as well as on the surface pollution. Pollution can result from the presence of particles which will be, to varying extents hygroscopic. For steel in indoor stores the critical humidity is usually given as 60%. Recently, however, it has been revealed that outdoor atmospheric corrosion of practical importance takes place only at a relative humidity which is higher than about 80%. This implies that the proportion of the time the relative humidity is 80% or more with the temperature during this time being above 0°C, can be described as the, so-called, *time of wetness*, during which corrosion takes place at a considerable rate. The time of wetness according to this definition can be calculated from meteorological data for temperature and relative humidity. With metal surfaces which are freely exposed out of doors, the time of wetness is naturally affected by rainfall and the drying conditions. Sensory devices have, however, been developed with the aid of which the time of wetness can be determined.

Barton, Bartonova and Beranek [8] have made the following estimations of the amount of water on metal surfaces under different conditions:

at the critical humidity	0.01 g/m^2
at 100% relative humidity	1 g/m^2
when the surface is covered with dew	10 g/m^2
when the surface is wet from rain	100g/m^2.

Figure 62 gives the relative humidity of the outdoor atmosphere at Uppsala in Sweden for different seasons of the year. From Fig. 63 it can be seen that the relative humidity can be reduced by raising the temperature and/or by the removal of water from the air (dehumidification).

Dry-air storage is often applied for temporary corrosion prevention of metals (see 6.8.3).

5.3.2 Effects of species in the moisture film

Moisture films on metal surfaces contain dissolved substances which affect corrosion; oxygen, sulphur oxides, including sulphate, (SO_x), nitrogen oxides, (NO_x), carbon dioxide, chloride and metal ions. Some of these have arrived by way of the atmosphere, others come from the corroding metal.

Oxygen

Oxygen usually takes part in the cathode reaction, which is necessary for the corrosion cells to function and for atmospheric corrosion to take place

$$\tfrac{1}{2}O_2+H_2O+2e^- \rightarrow 2OH^- \ .$$

The rate of oxygen supply from the air through the moisture film to the metal surface can therefore be of great importance. The supply of oxygen is most rapid, when the moisture film is thin and evaporation from the film is large, so that there will be good convection (movement) in the film. This means that the corrosion rate will be especially great when the surface is being alternately wetted and dried.

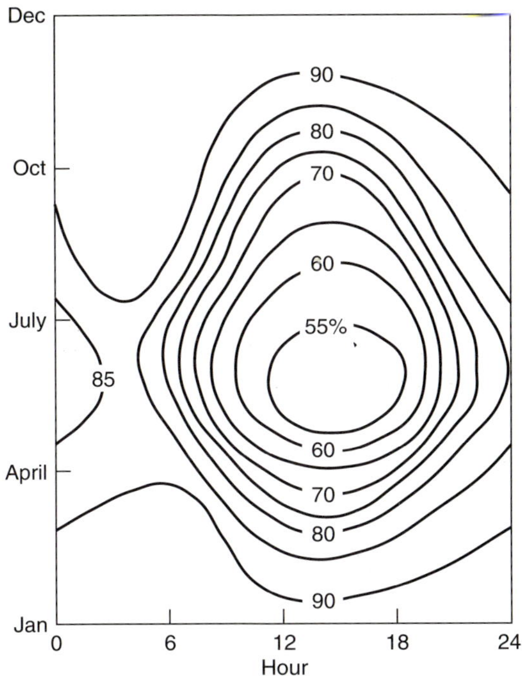

Fig. 62 Average of relative humidity during the day at different seasons of the year in Uppsala, 1866–1945 (Tropikutredningen, Robert Engström, Stockholm, 1968).

Sulphur oxides

Sulphur dioxide (SO_2) is of major importance for atmospheric corrosion. Adsorption of SO_2 onto the surface of a metal depends on the relative humidity and the presence of corrosion products. At 80% relative humidity or above, practically all SO_2 molecules which hit a rusty steel surface are absorbed. SO_2 oxidises to SO_3 in the atmosphere or in the moisture film on the metal surface and this, in turn, together with H_2O forms H_2SO_4. This acid reacts with the rust and is partly neutralised with the result that the moisture film becomes only weakly acidic with a pH value of about 4, whereas on surfaces which are not neutralising, e.g. painted surfaces or roofing felt, the pH value can be considerably lower.

Sulphur dioxide is formed predominantly by the combustion of coal and oil. Its dissemination depends, amongst other things, on the height of chimneys from which it is discharged. Thus, with low chimneys the pollution mainly affects the immediate surroundings, where it can significantly influence the atmospheric corrosion. But a few kilometres from the source of the emission the influence is small (Fig. 64). Higher chimneys allow a better spread of the pollution. The pollution which is spread

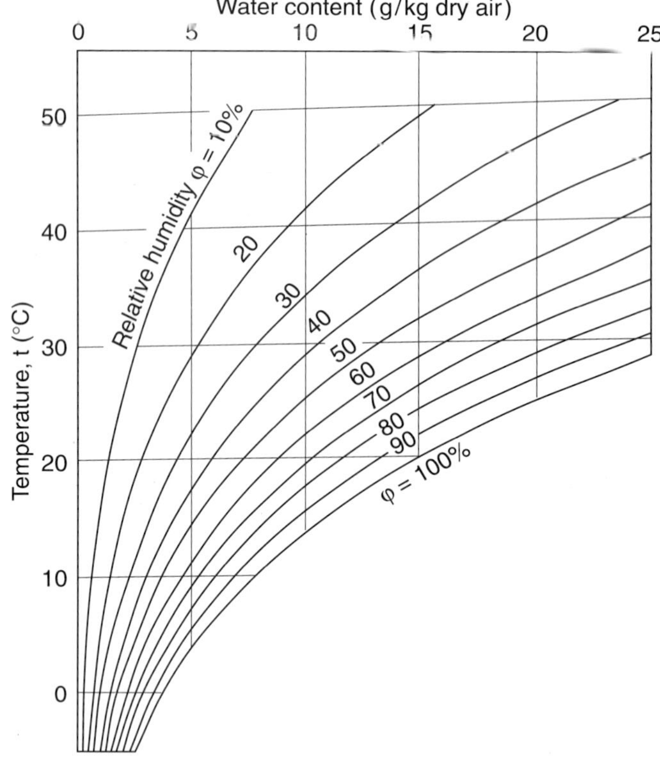

Fig. 63 Relationship between water content of the air, temperature and relative humidity (Carl Munters AB).

from a high chimney over a large area does not in general have any direct effect on atmospheric corrosion in the neighbourhood but can, when it returns to ground level, by dry deposition or as acid rain, lead to an increased corrosion rate of exposed metal surfaces, or to acidification of surface water, ground-water and soil. The corrosion of structures in these environments can thus be affected. These conditions are illustrated to some extent by how the rate of atmospheric corrosion for zinc has changed at the Swedish Corrosion Institute's testing station at Vanadislunden in Stockholm (Table 8). The corrosion rate increased during the latter half of the 1950s, obviously in connection with the introduction of oil-fired heating. By concentrating the combustion to district heating plants with high chimneys, in combination with the requirement of low-sulphur heating oil, the corrosion rate for zinc has over a period of 20 years been reduced to less than half of the value it had at the end of the 1950s.

Nitrogen oxides

Several nitrogen oxides occur in the outdoor atmosphere; N_2O, NO, NO_2 and N_2O_5. From corrosion point of view NO and NO_2 have attracted most attention. As these oxides are easily converted into each other by chemical reactions, they are often

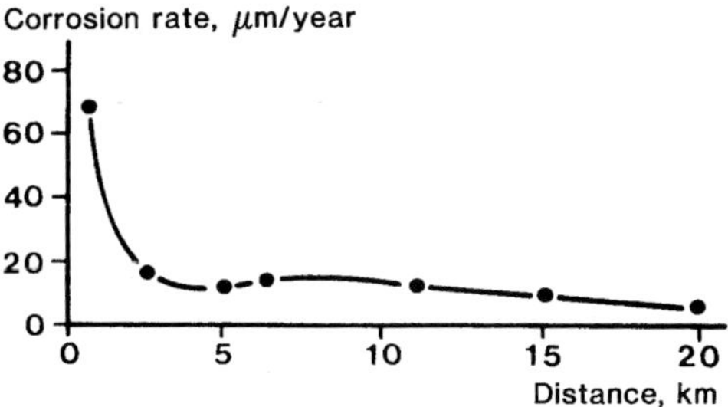

Fig. 64 The corrosion rate in carbon steel as a function of the distance from an emission source of air pollution (chimney) [9].

given the common designation NO_x. NO_2 is an oxidant which can cause corrosion of metals, being then itself reduced to NO. The two contaminants SO_2 and NO_2 together show remarkable synergism, increasing the corrosion rate considerably, e.g. on copper.

Table 8 — Corrosion rates of zinc on atmospheric exposure in the centre of Stockholm (Vanadis) [10]

Period	Corrosion rate (μm/year)
1938–1953	2.0
1958–1959	5.0
1958–1963	4.3
1967–1968	4.0
1967–1970	3.0
1975	2.9
1975–1976	2.3
1977–1978	2.1
1974–1978	2.0
1979–1980	1.7

NO_x may originate from, e.g. burning of fossil fuel, nitrification processes in soil and lightning discharges. In combustion engines NO is formed by reaction between oxygen and nitrogen in the air, due to the high temperature reached in the combustion cycle. By complex reactions of photochemical nature NO and hydrocarbons in the exhaust gases react with the formation of, e.g. NO_2 and ozone (O_3), which under extreme conditions may lead to the phenomenon of photochemical smog.

Chloride

Chloride is transported to metal surfaces either as droplets or as salt crystals in sea water spray. The deposition of chloride can therefore occur on metal surfaces which are exposed to wind from the sea. The droplets and salt crystals eventually fall to the ground, where trees, bushes and other obstacles can act as 'chloride filters'. The transport of chloride inland therefore varies considerably with the local conditions. Significant effects of corrosion as a consequence of the influence of chloride are usually limited to a narrow zone along the coastline and this is rarely more than a few kilometres wide.

Another chloride source is road salt ($CaCl_2$. $6H_2O$ and NaCl) spread on roads for dust control or for thawing of ice and snow. Spray and splashes of road water containing chloride will considerably increase the corrosivity of the road environment.

Dust and soot

Other air pollutants which can considerably influence atmospheric corrosion are dust and soot which are deposited on the metal surface. The effect depends to a large extent on the dust being able to retain moisture and acid remnants. Soot can also cause bimetallic corrosion since the carbon particles will act as effective cathodes for the oxygen reduction process.

5.3.3 The influence of temperature

Temperature also influences atmospheric corrosion. On the one hand a rise in temperature will increase the rate of corrosion as for that of other chemical reactions but, on the other hand, the time of wetness will be decreased thus counteracting corrosion. These two influences mean that the rate of atmospheric corrosion at a particular place reaches a maximum at a certain temperature. Below the freezing point of water the corrosion rate is generally negligible.

5.3.4 The classification of atmospheres with regards to corrosivity

In connection with corrosion it is usual to differentiate between the following main types of atmosphere:

— *rural atmosphere*, which has a relatively low level of pollution with a SO_2 deposition rate of less than 10 mg SO_2 per m^2/day and a chloride deposition which is less than 5 mg NaCl per m^2/day,

— *urban atmosphere*, which has relatively high levels of sulphur dioxide and soot, from cars, houses, district heating plants and industry; here the SO_2 deposition rate is 10-80 mg SO_2 per m^2/day

— *industrial atmospheres*, where the type of pollution can vary from industry to industry, but where the SO_2 deposition rate is often high and can be up to 200 mg per m^2/day; from the corrosion point of view industrial atmospheres often have a limited range and this is often restricted to the industrial area concerned, or even only to the immediate surroundings of the chimney from which the pollution is emitted, and

— *marine atmosphere*, which has a high concentration of chloride in the form of small salt crystals or drops from sea water spray; the annual average rate of deposition is usually between 5 and 500 mg NaCl per m^2/day, but can be as high as 1500 mg NaCl per m^2/day near the beach or over the sea, where the amount of sea water spray is greater.

The standard ISO 9223 is dealing with classification of the corrosivity of atmospheres. The standard defines in two equivalent ways five different corrosivity categories (C1 – C5) by specifying:

— the corrosion effects of the atmosphere on standard test specimens of carbon steel, zinc, copper and aluminium (see also ISO 9224),
— combinations of environmental data on time of wetness, pollution by sulphur-containing substances, and pollution by airborne salinity (see also ISO 9225).

On basis of the corrosivity category, selection of material and corrosion protection can be made.

5.3.5 The rate of atmospheric corrosion

The *instantaneous corrosion rate* varies considerably with the moisture conditions (Fig. 65). When a moisture film of sufficient thickness exists on the surface of a

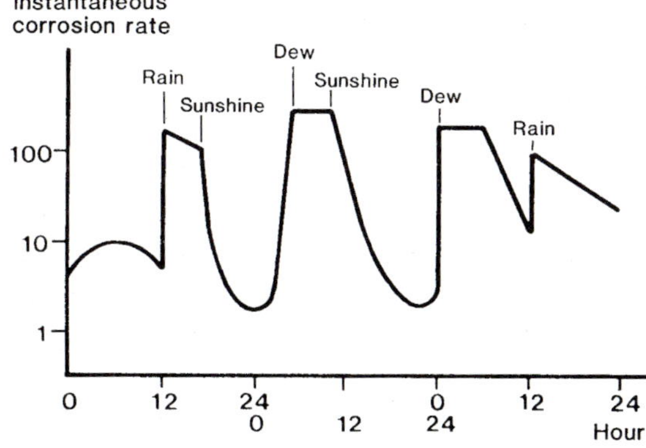

Fig. 65 The instantaneous corrosion rate (relative values) during some days of exposure [11].

metal, then the rate of corrosion will be comparatively high. When the surface is dry it will be several orders of magnitude lower. If the corrosion attack is recorded over a longer period, e.g. one year, then a continuous curve is usually obtained, as shown in Fig. 66. The tangent of the curve at any given point gives the *differential corrosion*

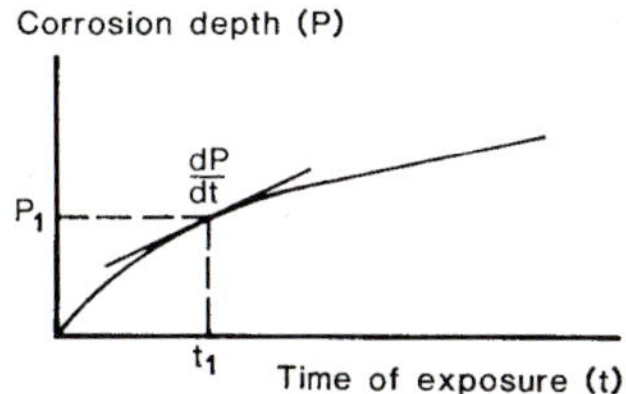

Fig. 66 Corrosion depth (P) as a function of exposure time (t); dP/dt = differential corrosion rate.

rate at that time. After a few years of exposure steady state conditions are usually reached, the curve straightens out, and the differential corrosion rate will be constant. In Table 9 the corrosion rates at steady state conditions in different types of atmosphere are given for the most commonly used metals.

Table 9 Approximate corrosion rates of steel, zinc, aluminium, and copper in different types of atmosphere

Type of atmosphere	Corrosion rate (μm/year)			
	Steel	Zinc	Aluminium*	Copper
Rural	5–10	0.5–1	<0.1	<1
Marine	10–30	0.5–2	0.4–0.6	1–2
Urban or industrial	10–60	1–10	~1	1–3

* On aluminium materials pitting is generally more important than uniform corrosion.

5.3.6 Atmospheric corrosion testing

The resistance of a material towards atmospheric corrosion is most reliably determined by *field testing* in the relevant atmosphere ISO 8565. This can be carried out by the long-term exposure of test pieces on testing racks (Fig. 67). The test pieces are usually mounted at 45° or 30° to the horizontal but can also be placed horizontally or vertically. The testing racks are generally positioned so that the front or upper face of the test piece is facing south. This is of special importance when testing painted material, where corrosion can be affected by sunlight. On well-equipped testing stations those climatic factors can be recorded which are most important for atmospheric corrosion, such as temperature, relative humidity, rate of deposition or concentration of SO_2, rate of deposition of Cl^- as well as the amount of rainfall and its pH value.

There can be differences in the exposure conditions between the test site and where the object is to be used. On a building for example, heat radiation can mean that the time of wetness can be different from that on the testing rack.

Since field studies usually require exposure times of several years, atmospheric corrosion is often studied by *accelerated testing* in some form of climatic chamber in the laboratory. But then it is important to ensure that the exposure conditions do not differ too much from those in practice so that the corrosion processes will be the same in the two cases (see 9).

5.4 DRY GASES

Oxygen, sulphur vapour and chlorine are examples of gases which in the dry condition can give rise to attack on metals. This means that the metal is oxidised and in the widest sense the attack is electrochemical in nature. Nevertheless, this corrosion can also be considered to be chemical corrosion, since it does not take place via electrode reactions in the conventional meaning of the expression.

Fig. 67 Site for corrosion testing in a marine atmosphere; Kvarnvik, Bohus Malmön (Mats Linder).

5.4.1 Reaction mechanism

Corrosion caused by an oxygen-containing atmosphere, leading to oxide formation on a metal surface, will be considered under this heading. The process can be divided into the following partial reactions:

$$Me \rightarrow Me^{n+} + ne^-$$
$$\tfrac{1}{2}O_2 + 2e^- \rightarrow O^{2-}$$
$$2Me^{n+} + nO^{2-} \rightarrow Me_2O_n \ .$$

These reactions take place in the presence of the surface coating of a previously formed oxide (Fig. 68). The formation of oxide according to the last formula can take place either at the Me_2O_n/air interface or at Me/Me_2O_n. The process consists of the following steps:

— the formation of Me^{n+} at Me/Me_2O_n,
— the formation of O^{2-} at Me_2O_n/air,
— transport of electrons (e^-) from Me/Me_2O_n to Me_2O_n/air,
— transport of Me^{n+} from Me/Me_2O_n to Me_2O_n/air (and/or of O^{2-} in the opposite direction), and
— reaction between Me^{n+} and O^{2-}.

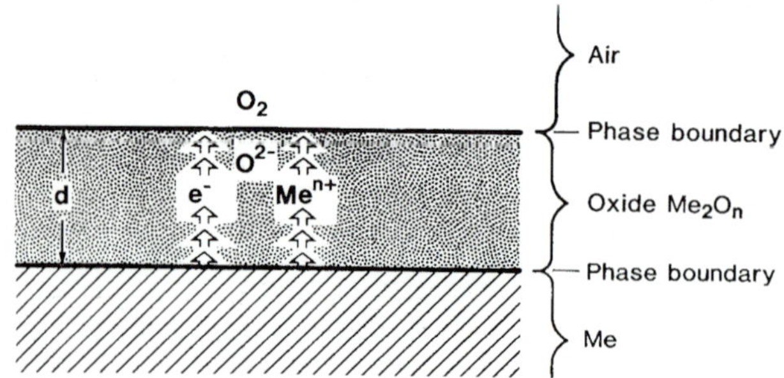

Fig. 68 Oxide formation on metal surface; *d* gives the thickness of the oxide.

The oxide coating is thus acting as a semiconductor. The transport of electrons takes place via the vacancies in the structure of the oxide crystals. Two types of semiconductors can be distinguished:

— *n-type* semiconductors, where the structure has an excess of Me^{n+} ions, which are compensated by the negatively charged mobile particles (electrons and ions), and

— *p-type* semiconductors, in which there is a deficiency of Me^{n+} ions resulting in positively charged particles being responsible for the transport of electricity.

As a rule it is the ion transport which is slowest and which therefore determines the rate of the whole process. In those cases where the structure has neither excess nor deficiency of Me^{n+} ions the surface coating will be an electrical insulator which hinders further film growth.

Usually, the growth rate of the oxide coating decreases with time (t), as the thickness (d) of the coating and the transport difficulties increase. The growth of the coating can obey different laws, depending on the reaction mechanism (Fig. 69). The

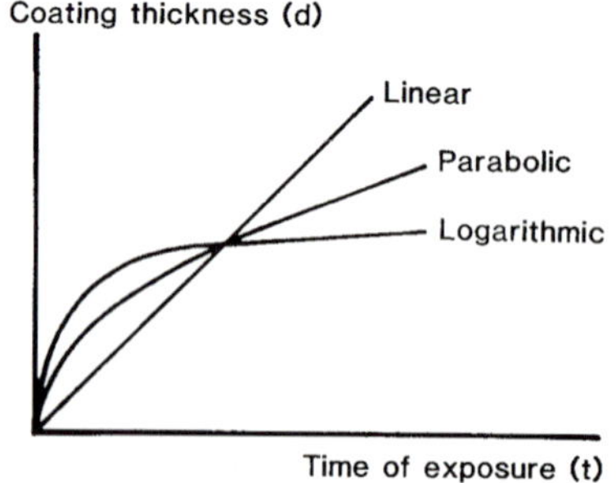

Fig. 69 Relationship between the thickness of the oxide layer (d) and time of exposure (t); examples of growth rate laws.

oxide/metal volume ratio is important. In those cases where the oxide formed has a smaller volume than the oxidised metal, then pores and cracks will appear in the oxide coating and further growth is favoured. In those cases where the oxide has a considerably larger volume than the metal then cracks in the oxide also appear with similar consequences. The most effective restraint on the further growth of the coating occurs when the oxide has a somewhat larger volume than the oxidised metal.

Seen from a technical point of view it is of course desirable that the growth of the oxide coating stops as soon as possible. At room temperature most metals soon get a thin, invisible oxide coating, the growth of which stops at a thickness of 10–50 Å (1 Å = 10^{-8}cm). At higher temperatures the oxide coating continues to grow on most metals, and to a considerable thickness on, for example, steel and copper so that a thick oxide scale is formed.

If a metal can exist in different oxidation states, the oxide coating that is formed will often consist of several layers with that richest in metal nearest the metal and that richest in oxygen nearest the air. In Fig. 70 the structure of an oxide coating formed

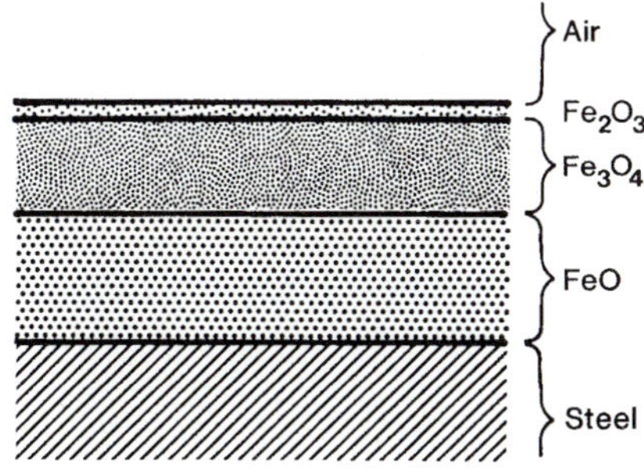

Fig. 70 Oxide coating on steel, formed in air above 570°C.

on steel at a temperature above 570°C is diagrammatically shown. Layers formed below 570°C do not have the FeO region. A similar situation exists with copper where at high temperatures a layer of Cu_2O occurs nearest the metal and CuO outermost.

Attack as a result of high temperature corrosion occurs for instance in gas turbines (Fig. 71).

5.4.2 Protective measures

High temperature oxidation can be counteracted by addition to the alloy of elements which tend to be selectively oxidised with the formation of a protective coating. For example, so-called heat-resisting steel has an addition of more than 12% chromium which leads to the formation of a thin, invisible layer of $FeO.Cr_2O_3$ and Cr_2O_3 at increased temperatures. This protects against further oxidation even at a temperature as high as 1000°C, if the chromium content is sufficiently high. Such steel is therefore used in high-temperature equipment, e.g. gas turbines. Under certain conditions, however, the protective properties of the oxide can be lost. This can occur if the surface is exposed to combustion gases with contaminants, e.g. vanadium oxide,

Fig. 71(a) Attack on a gas-turbine blade of nickel-base alloy caused by high temperature corrosion: appearance of the blade. (FFV Materialteknik).

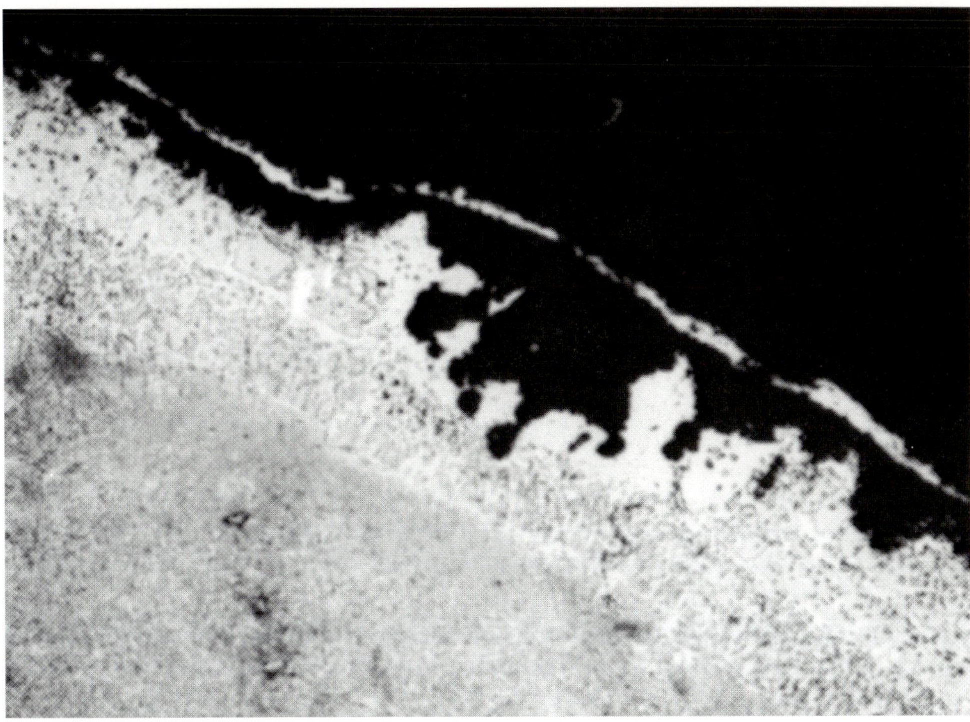

Fig. 71(b) Attack on a gas-turbine blade of nickel-base alloy caused by high temperature corrosion. Cross-section showing the attack at position indicated in Fig. 71(a) (FFV Materialteknik) (×400).

which can reduce the melting point of the protective coating. Oxidation can then continue at a high rate, and is then usually called *catastrophic oxidation*.

Protection by selective oxidation can also be attained by a thin alloyed layer at the surface, e.g. a diffusion layer of Al or Si, which gives a protective coating of Al_2O_3 or SiO_2 respectively.

High temperature oxidation can also be counteracted by the application of a coating, e.g.:

— aluminium, which at a high temperature can produce a thin, protective film of Al_2O_3; this coating can be used for steel up to 600°C,
— ceramics, e.g. ZrO_2, applied by flame spraying or plasma technique.

In order to avoid high temperature oxidation when heat treating reactive metals, a protective atmosphere consisting of nitrogen having low hydrogen, carbon dioxide and carbon monoxide contents, is often used.

5.5 THE HUMAN BODY

The human body provides very special corrosion environments to implants and restorations. Two types can be distinguished: the tissues of the body and the saliva in the oral cavity.

In orthopaedic surgery, metallic materials are often used for bone plates, screws and other fixation devices. In joint replacement systems a metal-plastics bearing arrangement is common, the plastics often being high-density or ultra high molecular weight polyethylene. Among metallic materials being used the following may be mentioned:

— stainless steels; generally steel of type EN 10088:1.4436 (former designation 316 L) or, when high mechanical strength is needed, a high nitrogen stainless steel; should better resistance to crevice corrosion be required, higher alloyed steels may be considered, e.g. EN 10088:1.4563,
— titanium materials; unalloyed titanium is commonly used; it has good corrosion resistance and also good compatibility with surrounding tissues; when higher mechanical strength is required, a titanium alloy with 6% aluminium and 4% vanadium is used, and
— cobalt-chromium alloys containing molybdenum or tungsten/nickel; these materials have good corrosion resistance and good mechanical strength as well.

In vivo the implants are exposed to the extracellular tissue fluid at about 37°C. This fluid is mainly a water solution containing sodium, chloride and bicarbonate ions, and in addition small concentrations of other inorganic and organic species. The salt content is of the order of 0.9 mass-%, and the fluid is isotonic with blood, i.e. it has the same osmotic pressure as blood. The pH value under normal conditions is about 7.4, but it may be lower after fracture trauma or a surgical operation. When the haematomas are being reabsorbed and the wounds are healing, it generally approaches the normal value. In crevices, however, low pH values may be persistent, e.g. at the prosthetic stem in a hip replacement.

The chloride-containing fluid in crevices having a low pH value may involve a risk of crevice corrosion in stainless steels susceptible to this type of corrosion. Further, implants may be affected by fretting corrosion, for example in the contact area between screw heads and plate holes, if movement occurs between screw and plate. The surrounding tissue may then be stained by corrosion products. In for example hip replacements, fracture by corrosion fatigue may take place due to bending and torsion. As the corrosion environment can not be changed, protective measures have to be focused on a suitable design, avoiding crevices as much as possible, and on an appropriate materials selection.

In the oral cavity various kinds of restoration components occur, e.g.:

— fillings replacing decayed parts in teeth; for anterior teeth aesthetic materials, e.g. composite resins containing a filler of glass; for posterior teeth with masticatory functions materials with higher mechanical strength, e.g. dental amalgam, the main constituents being mercury, silver and tin,

— crowns; for anterior teeth of porcelain, for posterior teeth gold alloy, e.g. gold-silver-copper, often covered with porcelain for aesthetic reasons,
— posts and pins for fixation of crowns; formerly made of brass or stainless steel which in many cases suffered from corrosion; nowadays generally made of unalloyed titanium or cobalt-chromium alloy, both with good corrosion resistance.
— bridgework, usually attached by soldering to crowns on adjacent teeth,
— removable partial denture, fabricated from a variety of noble and base metals as well as polymers, and
— orthodontics for realignment and positioning of teeth, often made of stainless steel.

The corrosive agent in the oral cavity is the saliva. It is largely a water solution of chlorides and phosphates but contains also organic acids, enzymes and gastric secretes. The pH value varies from person to person and also during the hours of the day for each individual. Teeth structures are also exposed to mechanical stress, especially the molars under mustication. Another matter to be taken into account is whether metallic contact between dissimilar metals occurs, which can cause bimetallic corrosion.

Apart from corrosion damage to posts and pins of brass or stainless steel, which were formerly commonly used, corrosion usually does not seriously affect metallic restorations. After relatively short exposure, passivation seems to take place which reduces the corrosion rate. For example surfaces of dental amalgam may be protected by a passivation layer containing species such as tin oxide/hydroxide, phosphates, hydroxide chlorides, mucopoly-saccaride etc. formed by reaction with the saliva. At scratches made in the passivation layer self-healing will take place. Corrosion effects on the environment seem to be of greater importance than the corrosion damage to the metal. Thus, metal ions released may cause staining or discolouration of teeth as well as of soft tissue. Their presence is sometimes recognised as a metallic taste. Further, it has been extensively discussed whether released ions, of for example nickel, can cause allergic reactions or lesions.

A more detailed review of metal corrosion in the human body including references to the comprehensive literature on this subject is given in: Shreir, L.L., Jarman, R.A. & Burstein, G.T. (eds). *Corrosion*, Vol. 1, 3rd ed., Butterworth & Heinemann, London, 1994, p. 2:155-180.

6
Corrosion protection

One can combat corrosion in many different ways, e.g.:

— by controlling the electrode potential so that the metal becomes immune or passive, i.e. by applying cathodic or anodic protection,
— by reducing the rate of corrosion with the aid of corrosion inhibitors added to the environment, or
— by applying an organic or inorganic protective coating.

6.1 CATHODIC PROTECTION

The corrosion rate of a metal surface in contact with an electrolyte solution is strongly dependent on the electrode potential. In most cases the corrosion rate can be reduced considerably by shifting the electrode potential to a lower value. This can usually be brought about by loading the surface of the object to be protected with a cathodic current, so that a negative polarisation is produced. This type of protection is therefore called cathodic protection. If the electrode potential is shifted so far that it corresponds to a point in the immunity domain of the potential-pH diagram the metal will be thermodynamically stable and corrosion cannot take place (Fig. 72). One can then refer to *complete cathodic protection*. If the shift in potential is less, the protection is said to be *incomplete*, but can nevertheless be of great practical value. For economic reasons, cathodic protection is often restricted to such parts of the structure, where the risk of corrosion is of particular importance, i.e. to so-called "hot spots".

6.1.1 Cathodic protection with impressed current

When the cathodic current to the surface of the object to be protected is supplied with the aid of an external current source (e.g. a transformer-rectifier unit) and an auxiliary anode (Fig. 73) then the protection is said to be of the *electrolytic type*.

In this case either soluble or inert anodes can be used. Soluble anodes can be made of steel (scrapped steel girders, rails etc.). The usual materials for the inert anodes are magnetite, silicon iron, graphite, lead or platinised titanium/niobium. For the protection of steel reinforcements in concrete there are flexible anodes of conductive plastics, consisting of a copper wire core surrounded by an electrically-

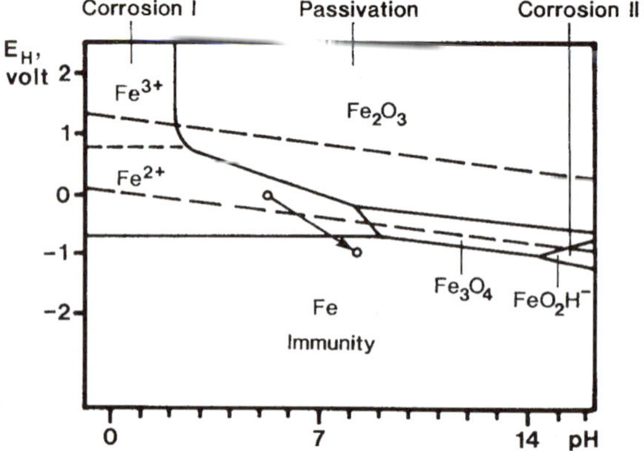

Fig. 72 Potential-pH diagram for Fe–$H_2$0 at 25°C; the arrow shows how, in principle, the electrode potential will fall as cathodic protection is applied, i.e. from a value within the corrosion domain to a value within the immunity domain with simultaneous development of alkaline conditions taking place at the surface of the protected object.

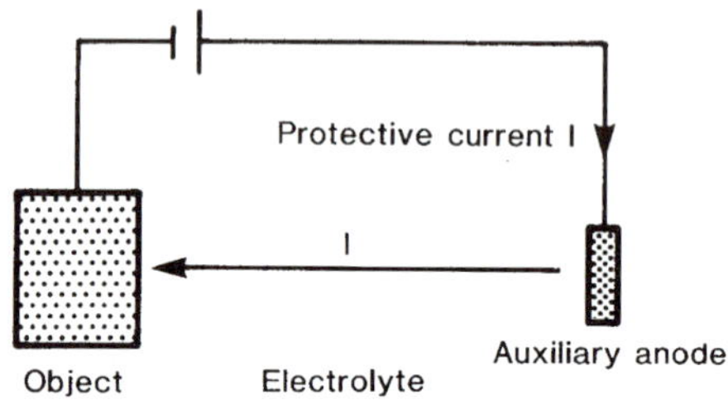

Fig. 73 — Cathodic protection with impressed current.

conducting plastics covering. The potential of the structure to be protected can be held at the desired value with the aid of a potentiostat and a permanent reference electrode which controls the flow of current from the rectifier. This technique is applied primarily when there are variations in the protective current demand. Cathodic protection with impressed current is used, for example, in the following cases:

— underground piping for water, oil and natural gas; it is usually economically advantageous to combine cathodic protection with an organic coating or by wrapping the pipe with tape,
— external protection of underground oil or gasoline tanks (Fig. 74),
— underground telecommunication cables with lead sheathing,
— steel piling in the ground,

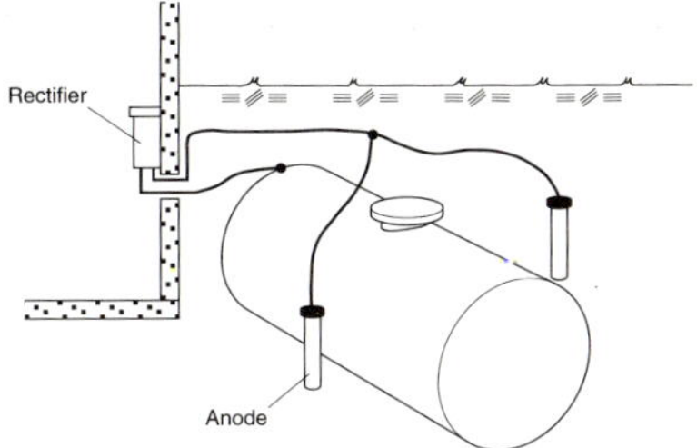

Fig. 74 Cathodic protection of underground steel tank with impressed current.

— steel sheet piling in harbours,
— steel reinforcement of concrete structures in air, using an anode system of, e.g. activated titanium net or conductive organic coating, covering the whole surface of the structure (for structures immersed in water external anodes can be used); protection with impressed current is not suitable for prestressed high strength steel reinforcement, due to the risk of hydrogen embrittlement,
— steel constructions in power plants, and
— internal protection of steel tanks for water or chemicals.

6.1.2 Cathodic protection with sacrificial anodes

The supply of cathodic current can also take place by electrical connection of the protected object to a less noble metal in the form of a so-called *sacrificial anode*, which is sometimes called a *galvanic anode* (Fig. 75) . The protection is then said to be

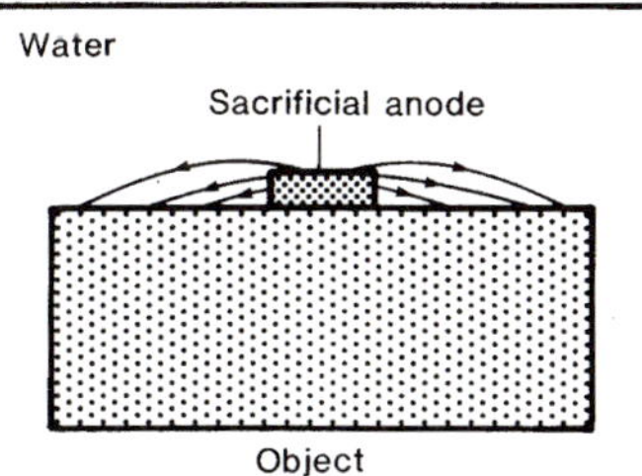

Fig. 75 Cathodic protection with a sacrificial anode fixed to the protected object.

of the *galvanic type*. The materials usually used in sacrificial anodes are magnesium alloys (MgAl6Zn3), zinc (99.99% Zn) or aluminium alloys (AlZn5). Iron is also used as a sacrificial anode for copper alloys. The sacrificial anodes are consumed or 'sacrificed', as a result of their protective action. Protection with sacrificial anodes is used for example in:

— the underwater parts of ships, especially near the propeller (Fig. 76); the latter is often made of copper alloy and could therefore cause bimetallic corrosion of the nearby parts of the hull,

Fig. 76 Cathodic protection of ship's stern with sacrificial anodes.

— internal protection of the tanks in oil tankers when the tanks are filled with sea water as ballast; in such a case magnesium anodes are unsuitable because this anode material can produce sparks if struck, thus giving the risk of explosion; zinc or aluminium anodes are more acceptable in this respect,
— the underwater parts of offshore platforms and pipelines on the bottom of the sea,
— steel reinforcement in concrete structures immersed in sea water,
— underground pipes and metal-sheathed cables (Fig. 77),
— the internal protection of oil and gasoline tanks made of steel, where corrosive water can collect at the bottom and cause pitting; a chain of sacrificial anodes is placed at the bottom of the tank; the anodes consist of magnesium bars which are fastened to a steel wire with a few centimetres of space between them,
— internal protection of hot-water tanks made of steel; in this case a centrally-placed magnesium anode can be suitable (providing the water has adequate conductivity) (Fig. 78), and

— in the water boxes of condensers (Fig. 79); there soft iron anodes are often used to protect the ends of the condenser tubes from erosion corrosion; zinc has been used to protect mild steel tube plates.

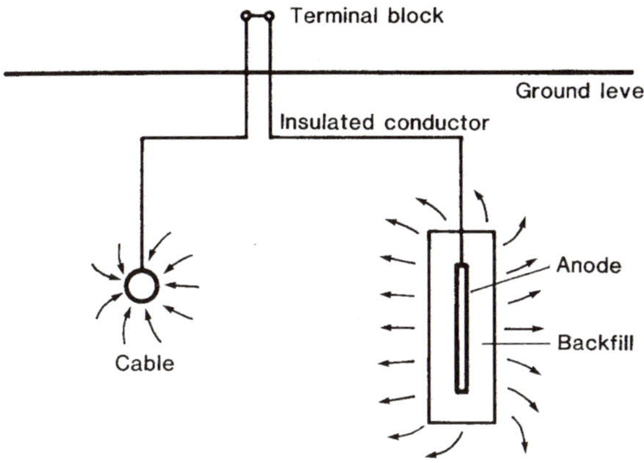

Fig. 77 Cathodic protection of underground cable with a sacrificial anode connected to the cable.

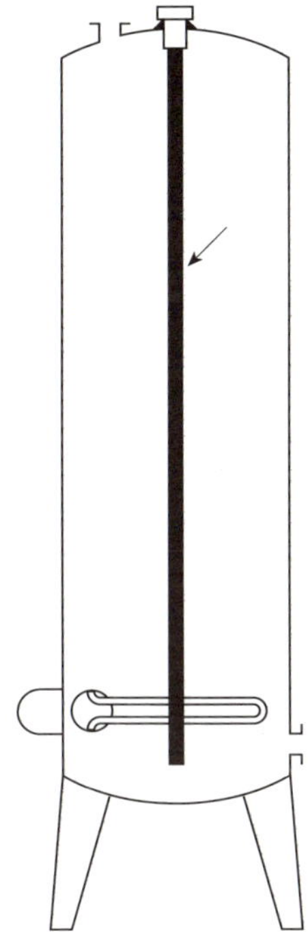

Fig. 78 Cathodic protection of a steel water heater with a centrally located magnesium anode.

Fig. 79 Cathodic protection in a water box of a condenser with sacrificial anodes.

6.1.3 Protection criteria

A decisive factor in the effectiveness of cathodic protection is the potential acquired by the protected object when cathodic current is applied. The potential is measured with respect to that of a reference electrode, for example in water a silver/silver chloride electrode, and in soil a copper/copper sulphate electrode. The measured value should be corrected for the IR drop which exists betwen the reference electrode and the protected object; techniques for this are available.

For cathodic protection of metal structures in soil and water the following protection potentials, E_p, are recommended, with respect to a copper/copper sulphate electrode:

iron and steel	$E_p \leqslant -0.85$ V,
lead	$-1.7 < E_p < -0.6$ V,
aluminium	$-1.2 < E_p < -0.9$ V, and
copper	$E_p < -0.2$ V.

At such values of the protection potential, corrosion is negligible under most conditions.

For iron and steel a protection potential of less than –0.95 V is recommended under certain conditions, such as in the presence of sulphate-reducing bacteria, at increased temperatures (e.g. 40 or 60°C) or at low pH values (e.g. pH 3.5).

For lead and aluminium a lower limit is given which should not be exceeded. The reason is that the hydroxide formation (alkalisation) which takes place at the cathode can lead to a dissolution of these metals with the formation of soluble plumbates and aluminates respectively.

Another criterion which is sometimes applied to cathodic protection is that the potential of the structure to be protected should be reduced by 0.100 V; thereby taking into account, and correcting for, the possible IR drop.

Cathodic protection of stainless steel is a special case, since the protection potential will lie within the passivity range (see 8.2) and not necessarily in the immunity zone. One can for example counteract pitting and crevice corrosion in stainless steel of the type AISI 304 in natural sea water by cathodic protection at a potential somewhat below –0.35 V with respect to a saturated calomel electrode [12].

6.1.4 Protective current

To bring about the desired reduction in potential for cathodic protection a certain protective current density is required on the protected structure. The requisite current density, usually given in mA m^{-2}, varies with the conditions and depends on the corrosive environment. For protection of uncoated steel the following current densities are generally required:

— in soil	10–100 mA m^{-2}
— in fresh water	20–50 mA m^{-2}
— in stagnant sea water	50–150 mA m^{-2}
— in flowing sea water	150–300 mA m^{-2}

If the surface has an organic (paint) coating then the current requirement is considerably less. Protective current is required only at pores and defects in the coating. For the protection of steel coated with glass-fibre reinforced bitumen, which is often the case in steel tanks, the current requirement is about 0.1–1 mA m^{-2}. With coatings of the polyethylene or epoxy type it is even lower say 0.01–0.1 mA m^{-2}. When cathodic protection is combined with organic coatings the spread of current over the protected structure is very good, and it is sufficient to have a few, well-located anodes.

The hydroxide formation which takes place on the surface of a protected structure causes, as already mentioned, a rise in the pH value. For combination with cathodic protection one must therefore choose coatings which are alkali resistant, e.g. bitumen, polyethylene or epoxy plastic. However, alkali formation can often lead to deposition of calcium carbonate on the protected structure. In time this can lead to a reduction in the current requirement and the cathodic protection of mild steel in sea water is an example of this. At very negative protection potentials (*overprotection*) hydrogen gas formation can take place on the protected surface. The hydrogen formation can in certain sensitive materials cause hydrogen embrittlement resulting in brittle fracture.

Anodes in soil are often surrounded with a *backfill*, which has good electrical conductivity in the vicinity of the anode, where the current density is greatest.

6.1.5 Interference

Cathodic protection with impressed current can in unfavourable conditions cause corrosion damage in nearby structures, e.g. pipes or lead-sheathed cables which are not connected to the protection circuit. This is called *interference* and takes place because the protection current may flow along the conducting pipe or cable in preference to the soil which is of higher resistance. This causes stray current corrosion (see 4.13) (Fig. 80). Such interference effects can be avoided by electrically connecting the structure in question to the protected object.

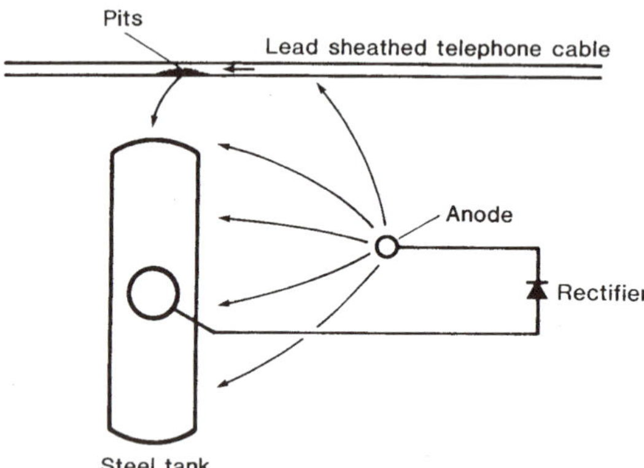

Fig. 80 Interference on a lead-sheathed cable caused by cathodic protection of a steel tank with impressed current.

Recommendations for the design of cathodic protection taking consideration of the avoidance of interference, are given in, e.g.:

— *Code of Practice for Cathodic Protection*, published by British Standards Institution as CP 1021:1973,

— *Control of External Corrosion on Underground or Submerged Metallic Piping*

Systems. Recommended Practice published in USA as NACE Standard RP-01-69 (1983 Revision),
— *Stray current and cathodic protection*, in the Swedish standard SEN 08 04 01, and
— *Cathodic Protection of Steel in Concrete,* Draft CEN Standard.

6.2 ANODIC PROTECTION AND PASSIVATION

An increase in electrode potential to more noble values converts certain metals from the *active* to the *passive state*. This applies, for example, to stainless steel in a sulphate solution. Passivation occurs when the so-called *passivation potential* is exceeded. When this takes place, the corrosion current is reduced by several orders of magnitude (see Fig. 118).

Anodic protection is based on passivation of the metal surface by the application of an anodic current. The anodic current induces positive polarisation, i.e. an increase in the potential of the metal. The current is such that the passivation potential is exceeded. If the potential is made too positive, then the region of passivation can be passed and pitting or so-called *transpassive corrosion* will occur. This type of corrosion protection is mostly applied in practice to stainless steel, where the iron is alloyed with chromium which has pronounced passivation characteristics. Anodic protection is also used on titanium and in certain applications on carbon steel.

The anodic current load can be provided with the aid of an external current source and a counter electrode in a manner similar to that used for cathodic protection, except that the current direction is in the opposite sense. Anodic protection of this type is used, for example, in vessels containing sulphuric acid.

However, it is more usually the case that the raising of the potential and the consequent passivation are effected with the aid of an oxidising agent in the electrolyte, e.g. HNO_3 or $K_2Cr_2O_7$. Even oxygen dissolved in the electrolyte is effective. In this case the efficiency can be raised with the aid of so-called microelectrodes on the metal surface. These consist of small particles of a metal, on which the oxygen can be reduced at low overpotential. Examples of such metals are platinum, palladium and copper. The resulting increase in cathodic current will be sufficient to move the potential into the passive region.

6.3 CORROSION INHIBITORS

A *corrosion inhibitor* is a substance which reduces the rate of corrosion when added to the corrosive environment in a suitable concentration, without the concentrations of the corrosive species present being changed significantly. An inhibitor is, as a rule, effective when present in small concentrations.

6.3.1 General mode of action

The action of an inhibitor is usually to produce a protective film on the metal surface. This can take the form of a very thin (monomolecular) layer of adsorbed inhibitor. Another alternative is that a protective film is formed by chemical reaction between the metal and the inhibitor, possibly with corrosion products being involved. A further possibility is that a thick coating (>100 nm) of the inhibitor forms on the metal surface. In such cases one speaks of film-forming inhibitors.

Inhibitors can be transported to the surface from, e.g.:

— a liquid corrosive environment, in which the inhibitor is present in dissolved or dispersed form,
— a corrosion preventing fluid with an inhibitor additive,
— an anti-rust paint with active pigment,
— the atmosphere in a package; in this case an inhibitor is required having a suitable vapour pressure, a so-called volatile corrosion inhibitor, or
— the metal to be protected; the inhibitor can be included as an alloying constituent.

Inhibitors are divided into anodic, cathodic and mixed types according to which reaction is impeded in the corrosion process.

6.3.2 Anodic inhibitors

These chiefly influence the anode reaction and the anodic polarisation curve (Fig. 81). Certain anodic inhibitors, e.g. chromate ions (CrO_4^{2-}) and nitrite ions (NO_2^-) as

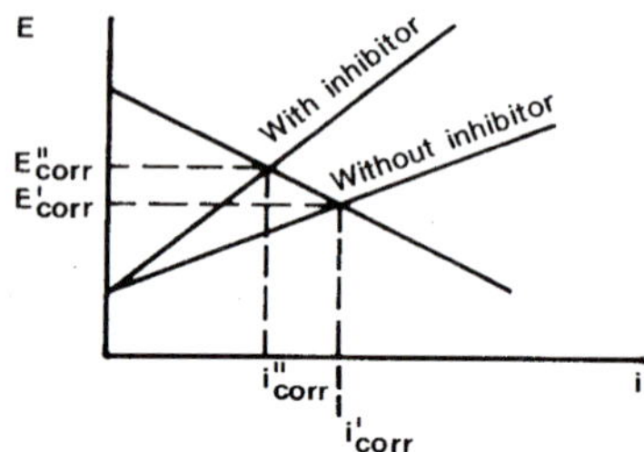

Fig. 81 Polarisation diagram, showing the action of an anodic inhibitor.

well as, in the presence of air, phosphate and molybdate, function by causing the formation of a protective (passivating) oxide layer on the metal surface. If the inhibitor concentration is too low, however, pores and defects can arise in the oxide layer, where accelerated corrosion can take place. These inhibitors are therefore called 'dangerous inhibitors'.

6.3.3 Cathodic inhibitors

This type of inhibitor chiefly influences the cathode reaction and the cathodic polarisation curve (Fig. 82).

Examples of cathodic inhibitors for steel are:

— zinc salts, e.g. $ZnSO_4$; their action depends on zinc hydroxide being precipitated at the cathode, where the pH value has increased, thus making the cathode reaction more difficult,

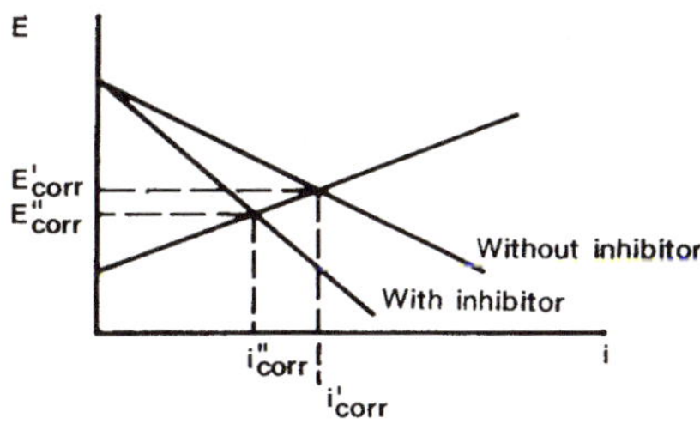

Fig. 82 Polarisation diagram, showing the action of a cathodic inhibitor.

— polyphosphates, e.g., sodium pyrophosphate ($Na_4P_2O_7$), sodium tripolyphosphate ($Na_5P_3O_{10}$) and sodium hexametaphosphate ($(NaPO_3)_6$); the inhibitor occurs in the colloidal form and gives, in the presence of two-valent metal ions such as Ca^{2+}, a protective coating on the metal surface, and

— phosphonates, which in the presence of two-valent metal ions and preferably in combination with a zinc salt, are effective as inhibitors.

Even with a low concentration of inhibitor, cathodic inhibitors provide some inhibition of the cathode reaction, and this counteracts the anode reaction. Therefore they are not 'dangerous' at concentrations which are too low for complete inhibition —in contrast to the situation found with correspondingly low concentrations of anodic inhibitors.

6.3.4 Mixed inhibitors

These influence both the anode and the cathode reactions to a larger or lesser extent (Fig. 83). Many organic inhibitors are of this type. Benzotriazole can be mentioned as an example which is widely used especially for copper and copper alloys.

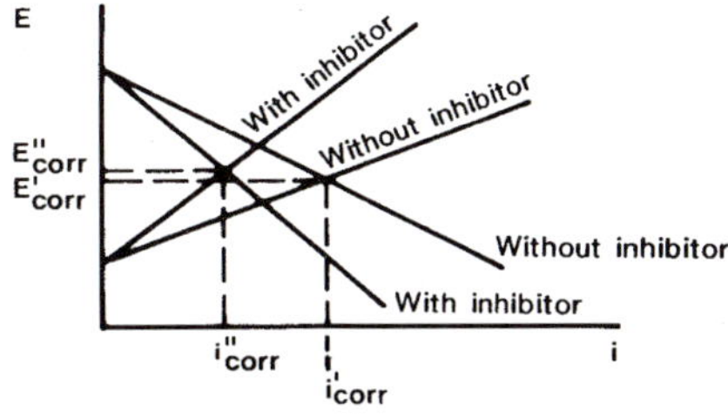

Fig. 83 Polarisation diagram, showing the action of a mixed inhibitor.

The character of many inhibitors is dependent on environmental factors, e.g. pH value and redox potential, so they are anodic under certain conditions and cathodic under others.

6.3.5 Application

In practice, mixtures of inhibitors are often used, so-called *formulations*, having a composition such that all the metals of which the structure is composed, are protected. When choosing an inhibitor one must, in addition to the corrosion-inhibiting effect, take consideration of factors concerning, amongst other things, industrial hygiene and environmental care in general.

The following applications of inhibitors can be mentioned:

— For water in open cooling systems, inhibitor formulations containing phosphonates and zinc salts are often used. These protect cast iron, steel, copper and aluminium materials.
— There are several metals used in the closed cooling systems of motor vehicles, e.g. cast-iron, aluminium alloys, steel, copper, brass, tin solder etc. In such systems the coolant will usually consist of a water-glycol (diol) mixture and temperatures will range from –47°C to >100°C, i.e. very corrosive conditions. An inhibitor formulation containing benzotriazole, silicate, borate, molybdate, nitrate and nitrite is an example for addition to glycol solutions.
— Pickling inhibitors can be added to pickling baths to prevent the dissolution of the metal being treated. Compounds of the thiourea type are used, for example, in sulphuric acid pickling baths for steel and silicates in alkaline cleaners for aluminium.
— The primer in anti-rust paint systems contains, as a rule, an inhibitor, a so-called active pigment. Important examples are red lead, zinc chromate or zinc phosphate. On contact with water species having inhibitive action are released.
— Slushing oils, corrosion-preventing fluids and greases normally have an inhibitor additive containing, for example, petroleum sulphonates.
— In transport containers and sealed packages volatile corrosion inhibitors are used, e.g. dicyclohexylammonium nitrite. This inhibitor can give protection to steel but can be corrosive to other metals, such as copper and lead. There are, however, other volatile corrosion inhibitors which protect these materials.
— Aluminium brass for condenser tubes has an addition of 0.02–0.04% As, which acts as a dezincification inhibitor. Low contents of arsenic, antimony or phosphorus have been shown to inhibit the dezincification of α-brass (see 8.4.8).

6.3.6 Evaluation

A large number of substances have been introduced as corrosion inhibitors, some of which in practice do not come up to expectations. It is therefore important to be able to evaluate inhibitors and determine the conditions under which they are effective. The efficiency of an inhibitor, I, can be expressed by the following equation:

$$I = \frac{V_0 - V_i}{V_0} \cdot 100\%,$$

where V_0 is the corrosion rate without inhibitor and V_i is the corrosion rate with inhibitor.

The corrosion rate is described by the mass loss of a test piece during a given exposure time.

The mode of action and efficiency of an inhibitor can also be studied by recording polarisation curves (Figs 81-83) or by polarisation resistance measurements (see 9.3). In recent years more advanced methods have become available for the study of corrosion inhibitors, e.g. electrochemical impedance spectroscopy and surface-enhanced Raman spectroscopy.

6.4 METAL COATINGS

Metal coatings are used to a considerable extent in corrosion protection. There are two main types which are characterised according to whether the coating metal is more or less noble than the substrate metal.

6.4.1 Corrosion protection properties

When *the coating metal is more noble than the substrate metal*, corrosion is accelerated in the pores, at sheared edges etc., where the substrate is uncovered (Fig. 84). If the coating is to provide protection it must be free from pores. The risk of all-

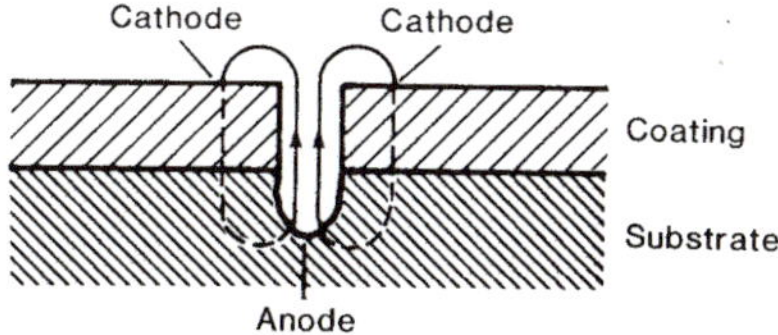

Fig. 84 Cross-section showing the conditions at the pores in a metal coating more noble than the substrate metal.

pervading pores is less the thicker the coating. The risk can be further reduced if the coating is produced electrolytically and applied in several successive layers. A coating of this type is chosen because it has better corrosion resistance than the substrate metal. An example is a nickel coating on steel.

A *coating which is less noble than the substrate metal* does not, on the other hand, need to be pore-free in order to provide protection, since it gives cathodic protection (Fig. 85). This works better the greater the conductivity of the electrolyte.

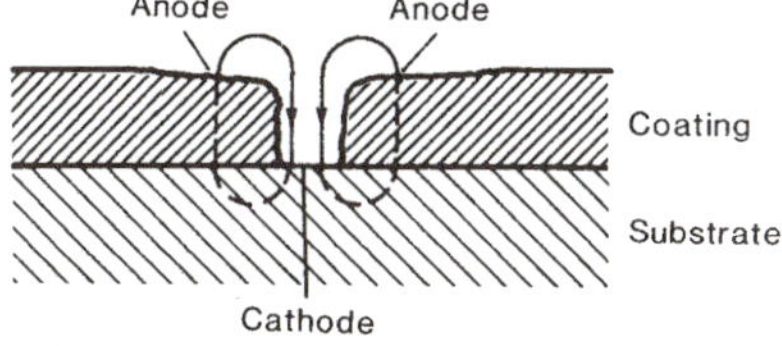

Fig. 85 Cross-section showing the conditions at the pores in a metal coating less noble than the substrate metal.

In an outdoor atmosphere, however, the cathodic protection is seldom effective over distances greater than about 1 mm. An example of this type of coating is zinc on steel. The protective action of the zinc is not, however, solely based on its ability to provide cathodic protection for the steel. Of greater importance is the fact that zinc is considerably more corrosion resistant than steel to outdoor atmospheres. The working life of a zinc coating is on the whole, proportional to the amount of zinc per unit area. The effectiveness of zinc is consequently more or less independent of how the coating is produced and whether or not the zinc is pure or present as a zinc-iron alloy. The corrosion rate of zinc coatings in different environments is given in Table 10. In a corrosive atmosphere the corrosion protection is often improved by painting the zinc coating.

Table 10 Corrosion rates of zinc coatings [13]

Environment	Approximate corrosion rate (μm/year)
Rural atmosphere	0.5–1
Urban or marine atmosphere	1–10
Marine atmosphere	0.5–2
Indoor atmosphere	<0.15
Fresh water	2–20
Sea water	10–25
Deionised water	50–200
Soil	~5*

*In certain soils much higher values occur.

It must be noted that zinc provides cathodic protection for steel only at temperatures below 50°C. At higher temperatures the electrode potential conditions can be reversed, so that zinc is more noble than steel. In hot-water tanks made of galvanised steel sheet the zinc coating can therefore give rise to pitting in the steel.

Metal coatings can be produced by several alternative methods. The choice of coating metal and the method of application depends upon several factors such as the metal to which it is to be applied, the size and shape of the object, the corrosivity of the environment, as well as the demands of the physical characteristics.

6.4.2 Electrodeposition

In this method the object is made the cathode in an electrolytic cell, in which the electrolyte contains the coating metal in the form of ions. The method is of great commercial importance. It is used for depositing pure metals such as zinc, copper, nickel, chromium, tin, gold and silver as well as alloys such as brass (copper-zinc) and zinc-nickel. *Electrogalvanising* is mainly used for producing thin zinc coatings. 'Chromium plating' for decorative purposes, e.g. on motor cars, implies the deposition of a composite layer consisting of 5–40 μm nickel for corrosion protection and on top of this 0.3–1.0 μm chromium to provide a shiny surface. Hard chromium plating is suitable for surfaces requiring good abrasion resistance and good frictional characteristics.

One must ensure good current distribution — *throwing power* — if the metal coating is to be even over the whole surface of the object. The current density usually tends to be too high at protruding edges and corners, while it is too low in cavities and at screened parts (Fig. 86). The current distribution can be improved by auxiliary anodes and current screens.

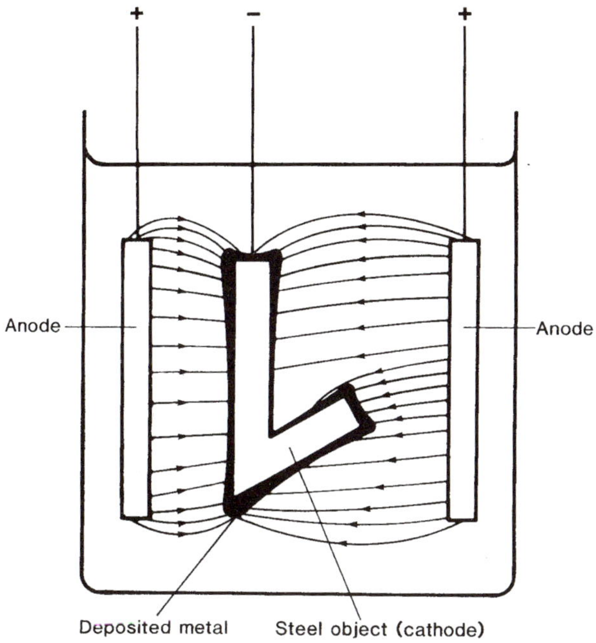

Fig. 86 Effect of uneven current distribution on electrodeposition of metals.

On electrodeposition of certain metals the formation of hydrogen gas can occur as a competing cathode reaction. Atomic hydrogen formed can diffuse in, and be absorbed by the substrate metal. In, for example, high strength steel, the result can be *hydrogen embrittlement*. The hydrogen can, however, be driven off by heat treatment thereby reducing the possibility of embrittlement.

6.4.3 Chemical plating

In this case the metal deposition takes place by displacement or chemical reduction. An example of metal plating by displacement is copper plating on steel according to the following formula:

$$Fe + Cu^{2+} \rightarrow Cu + Fe^{2+} \ .$$

The method is cheap but the coating is thin (*ca* 1 μm) as a rule, and porous and does not attach very well to the steel.

In electroless plating (chemical reduction) the deposition is carried out with the aid of a reducing agent which is added to the bath. A technically important example is nickel plating from an acidic hypophosphite bath according to the following overall formula:

$$Ni^{2+} + H_2PO_2^- + H_2O \rightarrow Ni + H_2PO_3^- + 2H^+.$$

With electroless nickel plating one can obtain a uniform thickness of the coating even on objects having crevices and complicated shapes. The deposition rate is practically constant and independent of the thickness of the coating. Chemical nickel plating with a hypophosphite bath gives an alloyed surface layer of nickel and phosphorus (2–13% phosphorus), in which the phosphorus content determines the characteristics of the coating in respect of hardness and ductility. Electroless plating is considerably more expensive than plating by electrodeposition.

6.4.4 Hot-dip metal coatings

In this process the object is dipped into a bath of the molten metal. Intermetallic compounds are formed at the metal-coating interface with the object so that an alloyed layer is produced in this region. The *hot-dipping* process is the most common method of zinc-plating steel when it is referred to as *hot-dip galvanising*. An alloyed layer is formed next to the steel and contains several iron-zinc phases; above this is a layer of zinc (Fig. 87). Hot-dip galvanising of steel sheet is done on continuous production lines (Fig. 88). In this way a 20–25 μm thick zinc coating is usually

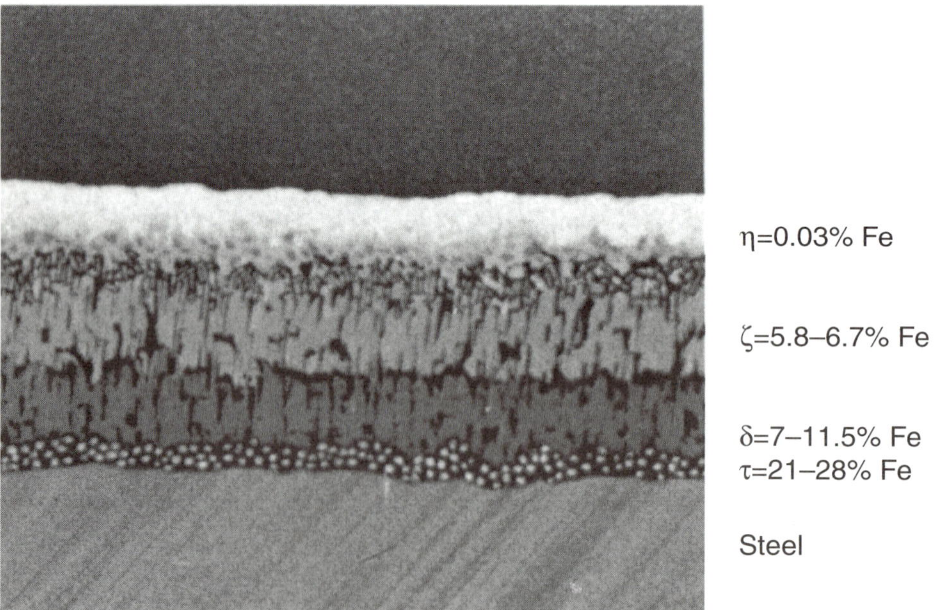

Fig. 87 Cross-section through a zinc coating formed on steel by hot-dipping; η, ζ, δ and τ are phases with different contents of iron; schematic diagram [13].

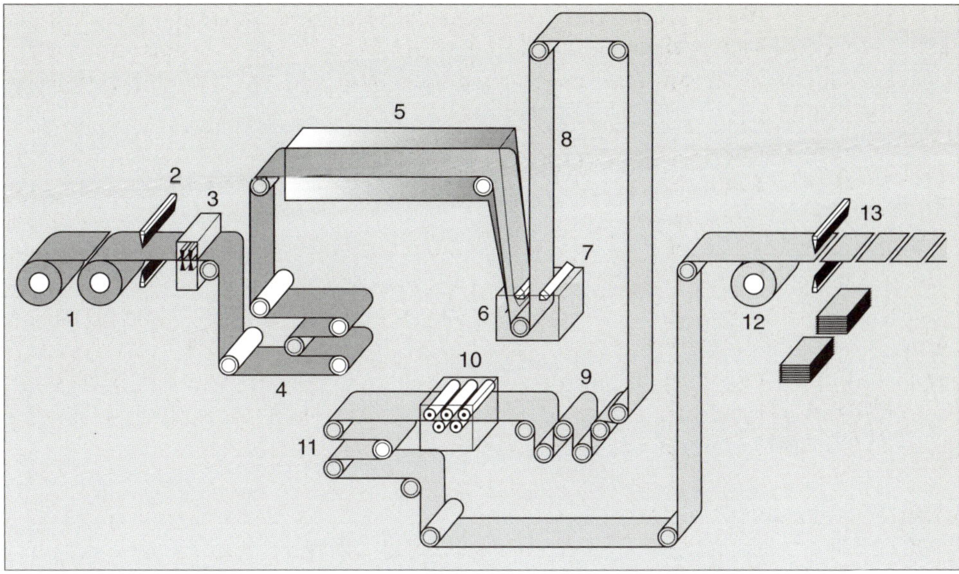

Fig. 88 Continuous hot-dip galvanising of sheet. 1 , cold-rolled sheet in coils; 2, cutting of the sheet; 3, welding; 4, magazine; 5, furnaces for surface oxidation, reduction and annealing; 6, zinc melt; 7, jet knives; 8, cooling; 9, levelling; 10, chromate passivation; 11, magazine; 12, coiling; 13, cutting [13].

produced. On hot-dip galvanising piece by piece one can obtain a maximum coating thickness of *ca* 70 μm on rimmed steel and considerably more on killed steel, at least 215 μm and in extreme cases 300–400 μm. Metal coating with tin, aluminium, aluminium-zinc alloys or lead is usually carried out by hot-dipping processes.

6.4.5 Metal spraying

Metal spraying is carried out with a spraying gun where the coating metal is melted or softened and projected in the form of drops towards the object.

The following types of metal spraying are in use (Fig. 89):

— *Flame spraying* of the coating metal with wire or powder, which is melted with an oxygen-acetylene flame. The metal is finely divided and transported to the workpiece by the flame and a flow of compressed air.

— *Arc spraying* of the coating metal with wire or powder which is melted in an electric arc between the wires. The metal is finely divided and transported to the workpiece by a powerful stream of compressed air.

— *Plasma spraying* of powder of the coating metal. The powder is melted by a plasma beam of for example, ionised argon, which is formed in an electric arc in the gun. The plasma beam has a very high temperature (*ca.* 15 000°C) and projects the molten metal drops with high velocity onto the workpiece. Plasma spraying is used primarily for coatings of materials with a high melting point, e.g. ceramics.

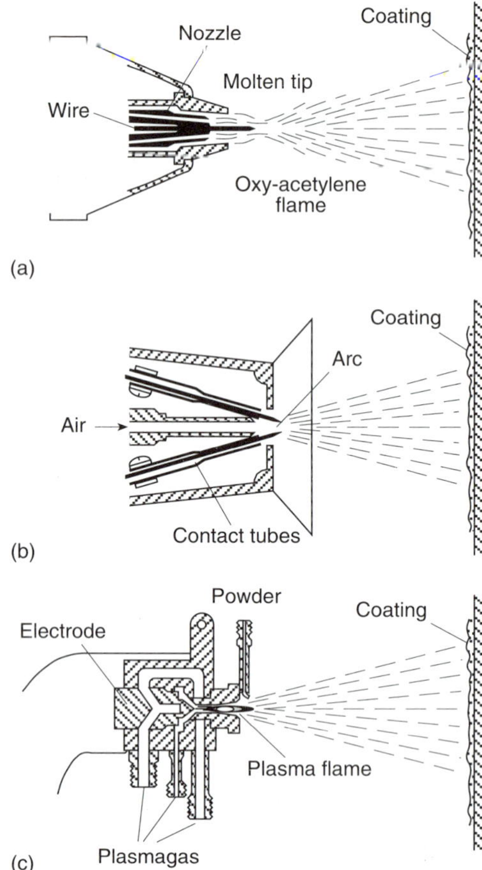

Fig. 89 Metal spraying. (a), flame spraying with wire; (b), arc spraying with wire; (c), plasma spraying with powder (Metco Sccandinavia AB).

By metal spraying one can produce coatings for corrosion protection having a thickness of 40–500 μm and in certain cases even more. On plasma spraying the molten metal is protected against oxidation and the drops strike the workpiece with considerable force. This leads to a low oxide content and low porosity (0.5–2%). With flame and arc spraying, however, there is no protection against oxidation and the impact power of the drops is lower. This leads to a higher proportion of oxide and greater porosity (3–7%). Because of the porosity the coating is sometimes sealed by painting.

Metal spraying is used for coatings of, for example, aluminium, zinc, stainless steel and lead. The method is suitable for coating large objects and for repairing damaged coatings, e.g. after welding.

6.4.6 Diffusion treatment

Diffusion treatment implies the changing of the surface zone of the substrate material by diffusion of a metal in from the surface. Diffusion treatment provides a coating of uniform thickness even with objects of complicated shape and the dimensions do not change significantly.

Of greatest practical importance is the diffusion treatment of steel with zinc. This process is also called *sherardising* and is used for motor-car parts, screws, nails, hinges and other small steel objects. On sherardising, the objects are packed into a container together with zinc powder or granules, sand and sometimes a halogen compound as activator. The container is sealed and placed in a furnace at 350–400°C for several hours. During the treatment the object acquires a zinc-rich surface zone, the thickness of which depends on the reaction time and can vary between 10 and 50 μm. Steel can also be diffusion treated with aluminium (*calorising*) or chromium (*chromising*).

6.4.7 Vacuum coating techniques

In *physical vapour deposition, PVD*, the coating metal is transformed to the vapour phase by volatilisation or cathode sputtering in a vacuum. The metal vapour is then transferred to the substrate metal and is allowed to condense there to form a coating. The temperature is usually 100–550°C. The method is used, amongst other things, for coating the reflectors of spotlights and headlights with aluminium.

In this connection *chemical vapour deposition, CVD*, should also be mentioned. In CVD treatment, the workpiece, in an environment at reduced pressure and at a relatively high temperature (800–1300°C), is exposed to a gas from which chemical reaction products deposit on the metal surface. CVD is used, amongst other things, in the production of surface coatings of TiC, TiN and Al_2O_3 on hard metal tools (Fig. 90).

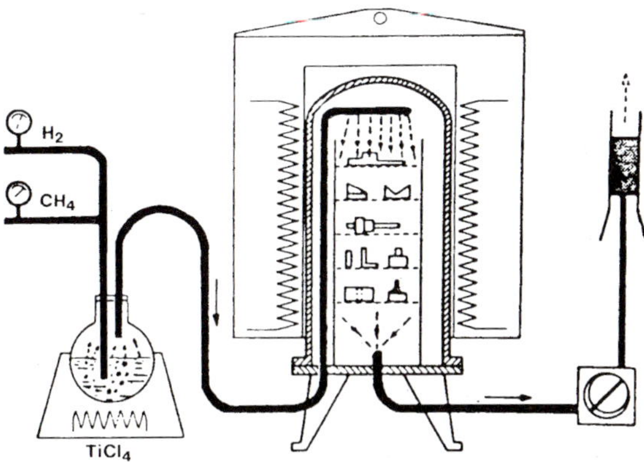

Fig. 90 CVD equipment for coating with TiC according to the formula $TiCl_4 + CH_4 \rightarrow TiC + 4HCl$.

6.4.8 Mechanical methods

There are several mechanical methods for metal coating, e.g.:

— *peen plating*; bulk quantities of goods are treated in a rotating drum containing glass beads together with powder of the coating metal suspended in water — as a result of the 'hammering' by the glass beads, the metal powder adheres to the surface of the components; the method is used for example for tin and zinc and can give coatings having a thickness of up to 75 μm,

— *metal cladding*; the coating metal is cold or hot rolled onto the substrate material,

— *explosive bonding*; a sheet of the coating metal and the substrate are welded together via an explosion,

— *extrusion*; the substrate material and the coating metal are extruded together, and

— *overlay welding* of the coating metal, e.g. high-alloy steel.

6.5 CHEMICAL CONVERSION COATINGS

In a chemical conversion treatment the metal reacts with the treatment agent so that a thin, difficultly-soluble coating is formed on the metal surface. Examples of chemical conversion treatment are phosphating and chromating.

6.5.1 Phosphating

Phosphating is used mostly on steel. One can distinguish between several variants of the process; zinc phosphating and iron phosphating or sodium/ammonium phosphating are those mostly used although manganese phosphating also occurs.

With *zinc phosphating* the steel surface is treated in a bath containing phosphoric acid, acid phosphate, zinc ions and certain additives, e.g. fluoride, nickel ions and organic compounds. Oxidation and dissolution of iron take place during the process and a small rise in pH value occurs near the surface. This results in the difficultly-soluble iron-zinc phosphate and zinc phosphate being deposited on the surface. The mass of the coating varies between 0.2 and 30 g/m^2 depending on the conditions. Phosphate coatings alone give only slight corrosion protection but together with subsequent oil impregnation good protection is achieved since the porous coating can absorb a great deal of oil. This method is used for protection of weapons, rock drills and certain motor-car parts. When used as pretreatment before anti-rust painting the phosphating effectively prevents rust from spreading under the paint at defects in the coating. This combination of phosphating and anti-rust painting is widely used for products made from cold-rolled steel sheet, e.g. car bodies. Zinc phosphating is used for galvanised steel which, however, must be free of the temporary protectives often used to combat white rust in storage. Zinc phosphating is also used on aluminium products.

Iron phosphating (sodium ammonium phosphating) is only used on steel. It is carried out in a bath containing monosodium or monoammonium phosphate at pH 4.0–5.5. The treatment is concluded by washing with a solution containing chromic acid and Cr^{3+} ions and in certain cases only the latter. The coating formed consists of ferrous phosphate ($Fe_3(PO_4)_2.8H_2O$), magnetite (Fe_3O_4) and certain iron-chromium compounds. It has a mass of 0.2–1.0 g/m^2 and the colour can be yellow/green, violet, blue or grey, according to the conditions. Iron phosphating is carried out in preparation for painting to improve the adhesion of the paint to the substrate. The method is applicable to sheet metal constructions to be used in moderately corrosive environments, e.g. household machines.

6.5.2 Chromating

Chromating is used, for example, on zinc, aluminium, magnesium and brass. The treatment is carried out using an aqueous solution of chromic acid or chromate with other additives, e.g. phosphoric acid and hydrofluoric acid are often present. A thin (0.1–2.0 g m^{-2}) chromate coating, having a green, yellow, black or pale blue colour, is formed on the surface and provides a degree of corrosion protection. Chromating is used a great deal on galvanised steel in order to give protection against white rust during transportation and storage. A considerable drawback arises from the fact that the handling of certain types of chromated material can produce skin complaints with some people. This arises as a result of contact with six-valent chromium. Another inconvenience is that the protective film is difficult to remove and can make subsequent painting difficult. Great efforts are being made to develop an effective protection against white rust which does not have the disadvantages of chromating.

Chromating is also widely used on aluminium, in preparation for painting, and as a decorative corrosion protection. Yellow chromating generally improves the adhesion of paint coatings on aluminium surfaces. A green chromate coating (without paint) is often seen in Sweden on aluminium roofs.

6.6 CORROSION-PREVENTING PAINTING

Corrosion-preventing painting implies that the metal surface is covered with a paint system which is able to prevent or delay the corrosion of the metal. Of prime importance is anti-rust painting, i.e. corrosion-preventing painting of steel.

Anti-rust painting includes the following steps:

— degreasing,
— the removal of mill scale, rust and other contaminants,
— possibly the application of etch primer,
— application of primer,
— possibly the application of an intermediate coating, and
— application of top paint.

6.6.1 Pretreatment

Degreasing can be carried out with an *alkaline cleaner* containing a surfactant or with an emulsifying agent and a grease solvent — *emulsion cleaner*. An alternative is degreasing with high-pressure steam — *steam degreasing*. Finally, degreasing may be conducted with a volatile *organic solvent*, e.g. white spirit, trichloroethylene or perchloroethylene. The degreasing treatment can be applied manually, by immersion, spraying or suspension in the vapour of the medium. Organic solvents have, however, lost some of their importance as a result of health hazards.

When anti-rust painting a steel surface it is important that *mill scale, rust and other contaminants* are removed if painting is to provide a lasting protection. When touching-up or repainting, previous coats, which have been damaged or peeled off from the substrate, must be removed. Cleaning can be carried out by *scraping* and *wire-brushing, grinding, blast-cleaning, pickling* (in industrial plants) or by *flame cleaning* followed by wire-brushing. Common abrasives for *dry blast-cleaning* are aluminium oxide, aluminium silicate, iron silicate or olivine sand as well as steel shot or grit. Quartz sand used to be the most common abrasive but may only be used under certain conditions now, as quartz sand can give rise to the disease silicosis. Dry-blasting with dry abrasive and compressed air is the most usual preparation method for large areas out of doors (Fig. 91). In industrial plants centrifugal blast-

Fig. 91 Blasting of steel pipe before anti-rust painting.

cleaning is carried out; in this a rapidly rotating wheel with blades flings abrasive against the steel surface. *Wet-blasting*, i.e. blast-cleaning with water added to the shot stream, has become more widely used recently, since, as distinct from dry-blasting, it is dust-free and even removes water-soluble surface contaminants such as chloride. In *hydro-blasting*, which is carried out with water under high pressure (e.g. 680 or 1700 bar), no abrasive is used.

6.6.2 Paint systems

In some cases *etch-primer* is applied. This is also called *'wash primer'*, and is designed to provide better adhesion for the subsequent paint layer. It contains polyvinylbutyral as binder, zinc chromate, iron oxide etc. as pigment as well as phosphoric acid as the etching component. Etch primer is applied in an amount equivalent to a dry film thickness of 2–7 μm. A modified etch primer, so-called *shop primer*, is used in order to provide blast-cleaned steel surfaces with protection during storage, transport and processing. It is often applied in large automated industrial plants.

Primers must have good adhesion to the metal surface and also provide it with corrosion protection. They therefore contain in most cases, besides the *binder*, an *active pigment* (a corrosion inhibitor). The following primers are the most common:

- red lead (Pb_3O_4) as an active pigment and linseed oil as binder; this primer is particularly useful for poorly prepared substrates but is poisonous and dries slowly; for reduction of the drying time the binder is often modified by the addition of an alkyd,
- zinc chromate as active pigment and alkyd resin as binder; this primer dries relatively quickly,
- zinc phosphate as active pigment and alkyd as binder, and
- zinc-rich paint, with zinc powder as 'pigment' and silicate or epoxy resin as binder; in this case zinc acts by providing cathodic protection to the substrate metal.

Primer is usually applied in one, sometimes two coats, having a dry film thickness of, in total 40–80 μm.

The *intermediate coat* is designed to provide good adhesion to the primer. It also contributes towards the building-up of a thick paint coating.

The *top paint* is designed to protect the primer against the effects of moisture, air and sunlight as well as give the surface its appearance. The chief components of the top paint are *pigment* and organic *binder*. The pigment should prevent light and water from reaching the substrate and provide the surface colour. Some examples of pigments are: titanium dioxide, iron oxides, aluminium flakes and barium sulphate. For outdoor purpose alkyd resin is generally used as binder. More chemically resistant binders are polyvinylchloride, chlorinated rubber, urethane and epoxy resins. The top paint is applied in one or two coats with a dry-film thickness of in total 40–240 μm.

The type of paint, the number of coats and the total thickness of the coats are chosen in relation to the corrosivity of the environment. It is important that the primer, the intermediate coat and the top paint are compatible. Directions are available for the choice of paint system in different cases, e.g.:

- *'Examples of approved rust protection systems'* in *The Regulations for Steel Constructions* (BSK) published by the National Swedish Board of Physical Planning and Building.

— *Code of practice for protective coating of iron and steel structures against corrosion*, in British Standard BS 5493
— *Steel Structures Painting Manual* published by the Steel Structures Painting Council (SSPC) in USA.

Painting of galvanised steel, aluminium and copper is carried out in a similar way as for steel, but the paint system is chosen to suit the respective metals.

6.6.3 Control

Anti-rust painting is an expensive surface treatment, the quality of which is very dependent on the preparation and application. Therefore careful control is required. The extent of the control varies with the nature of the object but the following points should be considered:

— the *cleanliness* of the surface after degreasing,
— the *efficiency of mechanical pretreatment*; with the aid of standards, ISO 8501/1–2, the amount of residual rust and mill scale can be determined by comparison with reference pictures,
— freedom from *dust* (ISO 8502/3) and *salts* (8502/1, 2, 5 and 6); methods for determination of these contaminations are described in the standards,
— *surface roughness*; can be determined by comparison with standardised gauges (ISO 8503/1-4),
— freedom from *risk of condensation* when the paint is being applied; the temperature of the steel ought to be a few degrees (3°C is usually specified) above the dew-point in the immediate surroundings; the method of determination is described in the standard ISO 8502/4; this requirement can be met by covering the object and heating or by dehumidification of the enclosed air,
— *thickness of the paint coating* (Fig. 92); the method of measurement is described in the standard SIS 18 41 60,
— *porosity of the paint coating,*
— *adhesion of the paint coating*; can be determined by pull-off testing according to the standard ISO 4623 (Fig. 93), or by cross-cut testing according to the standard SIS 18 41 72.

6.6.4 Paint coating breakdown

When an anti-rust-painted steel surface is exposed to a corrosive environment, e.g. an outdoor atmosphere, breakdown phenomena appear with time. Such phenomena can include rust stains, blisters and delamination, especially near scratches in the coating (Fig. 94). Under the influence of ultraviolet radiation in sunlight, oxidation or other kinds of breakdown of the organic binder will take place. This breakdown leads to pigment grains being released and forming a loose powder on the surface. The process is called *chalking*. Top paints have different resistances to chalking (Fig. 95).

The condition of the steel surface in respect of the occurrence of rust stains can be graded according to a six-level *scale of degree of rusting* for anti-rust paints (SS 18 42 03, ISO 46 28). The rust grade Ri O means, that on inspection no rust can be discovered on the painted surface, while rust grade Ri 5 means, that 40–50% of the

Fig. 92 Measurement of the thickness of a paint coating on steel with the aid of a transistorised dry-film meter.

surface is covered with rust. The grade increases successively during exposure. When the grade has reached about Ri 4, it is generally considered time for repainting.

6.6.5 Paints without organic solvent

In conventional paints the binder is dissolved in an organic solvent, e.g. white spirit. The solvent can however lead to problems from the hygiene point of view, since unhealthy vapours can form on evaporation. For this reason considerable development work is being carried out in order to produce environmentally safer paints, e.g. *water-based paints* with water as 'solvent', *solvent-free paints*, where the binder is itself liquid and *powder paints* which are applied in powder form. In this develop-

Fig. 93 Testing the adhesion of a paint coating with a Säberg instrument.

ment work it is important that the corrosion protection ability of the new paints is confirmed by field tests before the paints are marketed for large-scale application.

6.7 COATINGS OF PLASTICS OR RUBBER

With the aid of coatings of plastic or rubber one can protect metal surfaces which are exposed to extremely corrosive chemicals. This method of protection is used, for example, in storage containers for chemicals, reaction vessels, electrolytic cells, pickling tanks, pipes, pumps, fans and suchlike. Plastic coatings are also used on sheets of galvanised steel and aluminium intended for building purposes.

6.7.1 Coating materials

Three main types of coating materials can be distinguished:

— *thermosetting plastics*, such as phenolic, epoxy, and polyester plastics,
— *thermoplastics*, such as ethylene, propylene, amide, vinyl, vinylidene chloride and tetrafluoroethylene plastics, and
— *rubbers*, such as natural, butyl, chloroprene, nitrile and hard rubber.

The various types of plastics and rubbers differ considerably with regard to applicability, adhesion, chemical resistance as well as resistance to mechanical and

(a)

(b)

Fig. 94 Coating behaviour at scratches: (a), delamination of coating; (b), no delamination due to suitable primer treatment.

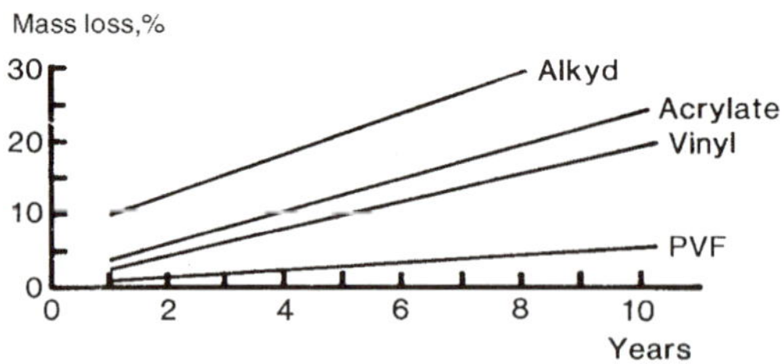

Fig. 95 Mass loss of paint coatings with different types of binder on exposure to the atmosphere in Florida [14].

thermal stresses. The choice of the most suitable material can best be made with the aid of handbooks with tabulated data on this subject.

6.7.2 Application of coatings

In certain cases coatings can be produced by the application of a solution or dispersion (small particles or drops suspended in a liquid) in a similar manner as with painting, i.e. by application through brushing, dipping or spraying. Another possibility is to heat the object and bring it in contact with a powder of the coating material — although this applies only to thermoplastics. This treatment can be carried out in a so-called fluidised bed or by spraying. By using these methods it is possible to obtain coating thicknesses of 0.2–2 mm. Thicker coatings, 1–6 mm, can be obtained by the use of adhesives to attach films or tiles, following careful cleaning of the metal surface, e.g. by blasting treatments. Glass-fibre-reinforced plastics (GRP) coatings are produced by the application of glass cloth or chopped glass-fibre together with a solution of resin.

6.7.3 Types of coating failure

The coating must adhere well to the substrate in order to fulfill its function. Further, it must have satisfactory chemical resistance as well as sufficient durability against knocks and bumps, wear and thermal stresses. Types of damage which can result from chemical processes are:

— *hydrolysis* by the action of water, especially at raised temperatures; this leads to a reduction in strength or complete destruction,
— the *splitting off of hydrochloric acid* by the action of UV-radiation, e.g. sunlight; this leads to discoloration and a reduction in strength,
— *oxidation* by the action of oxygen in the atmosphere or strong oxidising agents; this leads to a breakdown which can cause embrittlement or the formation of a soft, sticky material,
— *liquid absorption*; this leads to swelling, and
— *stress corrosion cracking* by the combined action of tensile stress and chemicals; this leads to cracking usually through the whole thickness of the material.

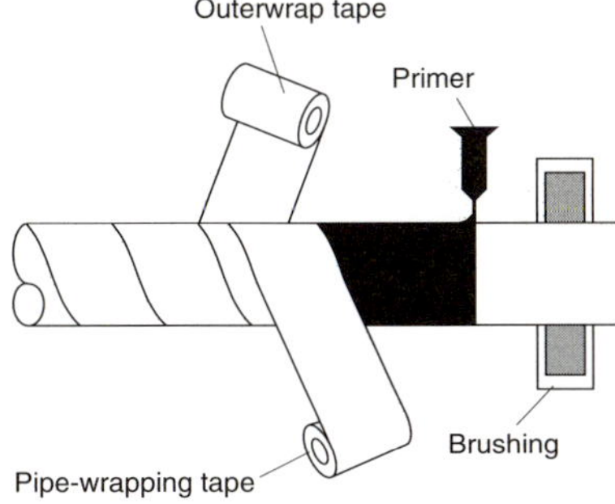

Fig. 96 Corrosion protection of a pipe by tape-wrapping (Nitto Scandinavia).

6.7.4 Tape

Tape for corrosion protection can be used, especially for piping, e.g. underground piping. Before the tape is applied the pipe surface must be cleaned of oil, previous coatings, rust and foreign matter. Then a primer is applied in order to provide good adhesion of the corrosion protection tape. The tape consists of an approximately 0.5 mm thick film of PVC or polyethylene plastic. It is often complemented with an outer tape for mechanical protection (Fig. 96). The application of the different components can be done by hand but can be carried out on a large scale by special winding machines. At joints, protection is afforded with the aid of sleeves of shrinkable plastic, which on heating contract to give a tight coating. Tape wrapping is often combined with cathodic protection which will prevent corrosion at pores or gaps in the wrapping that may arise during application and installation.

6.8 TEMPORARY CORROSION PREVENTION

In certain instances metal objects need only be protected from corrosion during a limited period. This will be the case in the protection during storage or transportation of, for example, machines, engines, gearboxes, instruments, tools and utensils as well as semi-manufactured products, such as metal strip, sheet, wire and pipe. Such protection is called *temporary corrosion prevention* and is usually achieved with the aid of film-forming agents or volatile corrosion inhibitors or by dry-air storage.

6.8.1 Film-forming agents

Film-forming agents provide the metal surface with a coating which excludes moisture and in this way hinders corrosion, for a while at least. The film-forming agents usually contain a corrosion inhibitor (see 6.3) which contributes towards arresting corrosion. Film-forming agents are applied by dipping, spraying or brushing (Fig. 97). The following variants can be distinguished:

— *Water-based corrosion-preventing agents* are emulsified in water. When the water has evaporated a thin, often oily, film remains. This can, however, be re-emulsified and washed away if rain falls on the surface.

Fig. 97 Application of corrosion-preventing fluid by 'airless' spraying (Dinol Svenska AB).

— *Corrosion-preventing oils* consist of a mineral oil with an addition of corrosion inhibitor. The oil provides a non-drying film, which can easily be removed with, for example, an organic solvent.
— *Corrosion-preventing fluids* are composed of grease, oil, wax or resin together with corrosion inhibitor dissolved or dispersed in an organic solvent. When the solvent evaporates a protective film is formed, having a thickness of <1 μm to >200 μm. The film can, depending on the film-forming substances, be greasy, oily, wax-like or resemble paint.
— *Corrosion-preventing greases* also have an addition of corrosion inhibitor. They form a soft, fatty-like film which gives very good corrosion protection.
— *Hot-dipping agents* are applied by dipping into a melt of vaseline (or similar petroleum product), wax etc. A thick solid film is obtained providing excellent corrosion protection.
— *Polymer films* include thin, dry, strongly adhesive plastics films, and strippable plastics films deposited from solvents or applied by hot dipping. The hot-dip coating is thick, strong, dry or oily and easily removed.

6.8.2 Volatile corrosion inhibitors

Volatile corrosion inhibitors (VCI), are transferred to the metal surface in the gaseous state. Volatile corrosion inhibitors are suitable for protection in closed spaces, e.g. in packages or closed containers. The inhibitor can be applied either in the wrapping paper or in the form of powder, tablets or capsules enclosed in the package. Many volatile corrosion inhibitors found on the market have dicyclohexylammonium nitrite as the chief ingredient. Both this and other related inhibitors provide good corrosion protection for steel but can cause corrosion of lead and copper. Inhibitors have however been developed which are suitable for metals other than steel and even some which can protect several metals. A limitation for many volatile corrosion inhibitors is that the presence of water in liquid form reduces their protective ability. A major advantage is that the protected products can be used or further processed directly without prior cleaning. The saturated vapour pressure of a VCI is an important property. If this is too high then the inhibitor will tend to escape from the container through the slightest defect in the sealing; if too low the 'throwing power' will be insufficient to provide protection over any significant distance.

6.8.3 Dry-air storage

With *dry-air storage* the relative humidity of the air is reduced in the storage space to a level below the *critical humidity*. Under such conditions the corrosion rate of the metal is considerably reduced. For steel the critical humidity is about 60%. One can reduce the relative humidity in the storage space by *increasing the temperature* to a few degrees above the temperature out of doors. This works well during the winter but is difficult to bring about during the summer. During this period the risk of corrosion in stored goods is therefore great. The relative humidity can also be reduced with the aid of a *dehumidifier* which can be set to desired relative humidity, e.g. max 45–50%. Dehumidification can be used for, e.g. storerooms, garages, equipment in 'mothballs', ships' holds and containers for the transport of goods (Figs 98 and 99).

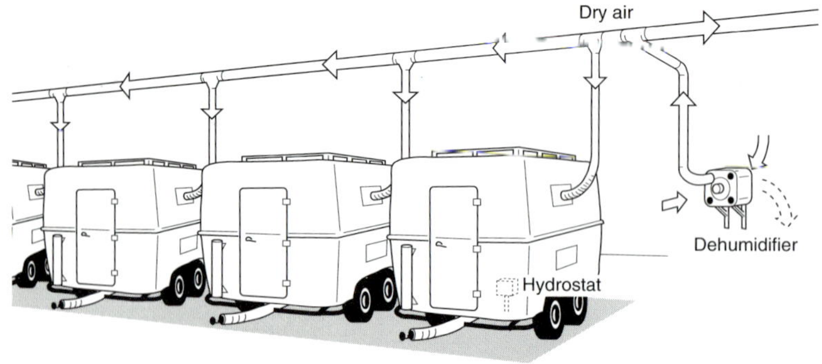

Fig. 98 Dehumidification of air to long-term parked vehicles (Carl Munters AB).

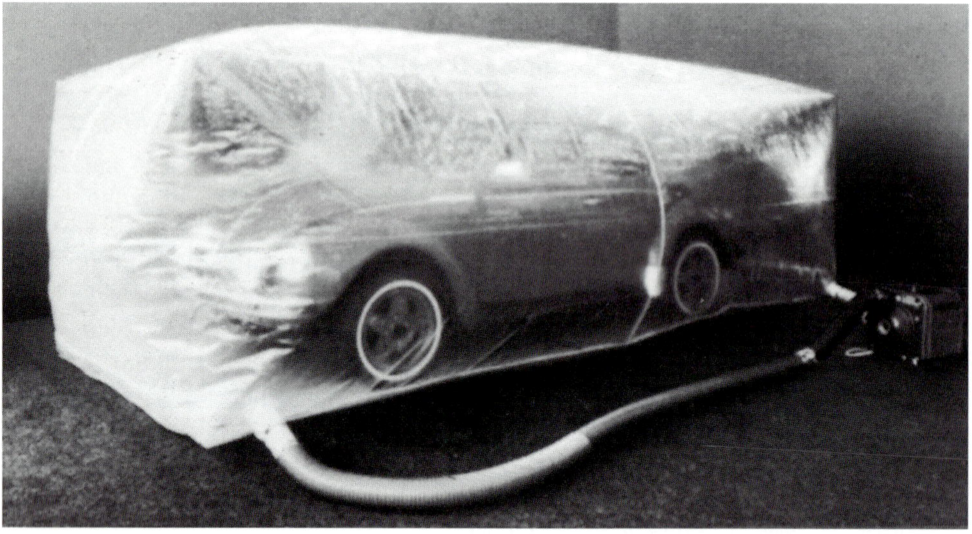

Fig. 99 Temporary corrosion protection of a motor-car by dry-air storage in a plastic envelope (Carl Munters AB).

The 'storage space' is often a *package*. Then the packaging material should be impenetrable for water even in the vapour state. Such packaging material is for example, polyethylene foil, bitumen-impregnated paper, waxed paper as well as paper laminated with aluminium foil. In order that the water vapour barrier will be effective it is required that joints are sealed in a suitable way. A *drying agent* is included when the package is sealed, e.g. silica gel.

Dry-air technique is also being applied for internal corrosion protection of hollow constructions, e.g. steel bridges, motor vehicles and aeroplane engines.

7
Corrosion prevention by design

The design of a construction contributes to a large extent towards its resistance to corrosion and thereby to its service life.

7.1 THE THICKNESS OF THE MATERIAL

In many cases the requirements for tensile strength are decisive for the dimensioning of a construction. In cases where there is a risk of *stress corrosion cracking* (see 4.11), one must ensure that the tensile stresses do not exceed the maximum value which from the viewpoint of stress corrosion cracking is acceptable for the alloy in question. With alternating loading one should ensure that the endurance limit is not exceeded otherwise damage from *fatigue* or *corrosion fatigue* can occur (see 4.11). The risk of cracking from stress corrosion, fatigue or corrosion fatigue is greatest where there is a concentration of mechanical stress, e.g. at notches and small holes, as well as at places where there are abrupt changes in shape. Such irregularities should be taken into account by the inclusion of a special form factor when the strength of a construction is being calculated for dimensioning. In welded constructions it should be remembered that the structural strength as well as the resistance to stress corrosion cracking, fatigue and corrosion fatigue are in many cases reduced at or near the welds.

Sometimes dimensioning is done with a *corrosion allowance*. This implies that the material thickness is increased by an amount which will compensate for expected corrosion. Such an increase in the material thickness can be limited to those parts of the construction where the corrosion rate is expected to be especially great, e.g. at the boundary zone between liquid and air (see 5.1.1).

7.2 'POCKETS' WITH ACCUMULATED WATER OR DIRT

In places where water or dirt remain, the risk for corrosion is especially great. In designing a construction care should be taken to ensure that there are no 'pockets'; where water or dirt can accumulate. Figure 100 shows how certain structural members ought to be designed so that the collection of water and dirt is avoided. The problem also exists in certain types of liquid containers (Fig. 101). As a consequence

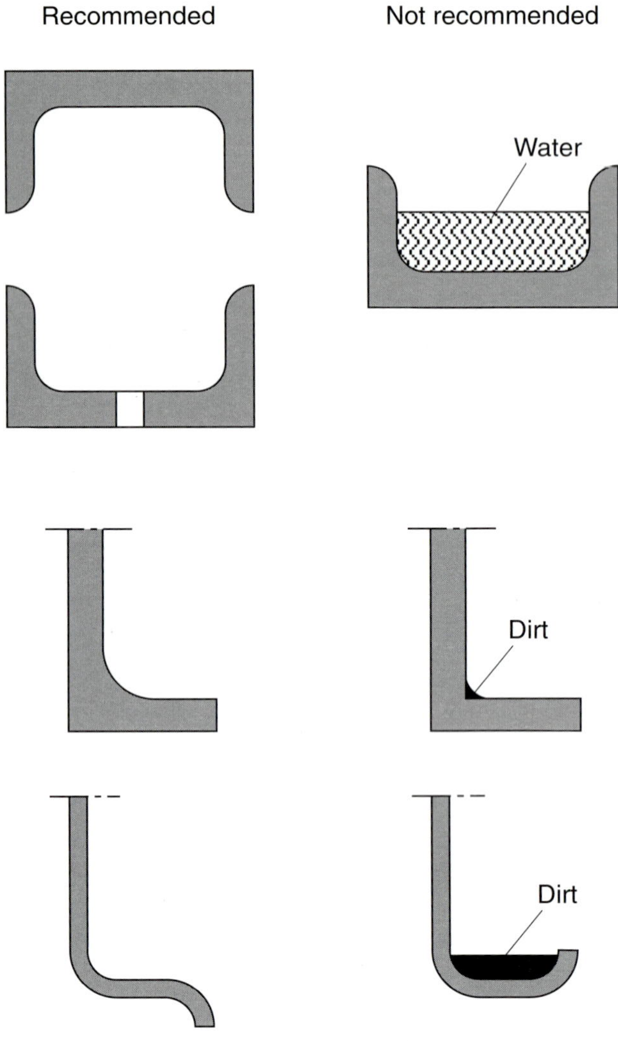

Fig. 100 Avoid such design that 'pockets' collecting water or dirt arise.

of the unfavourable design these cannot be completely emptied — at least not without an unreasonable amount of work. When these containers are left with liquid remaining in the 'unemptied' pockets for long periods of time, corrosion will take place in these sites. Figure 101 shows examples of good and bad design.

As a further example, metal linings of walls of buildings can be mentioned. Here steps should be taken to ensure adequate ventilation and drainage inside the wall, Fig. 102, so that 'pockets' of accumulated water do not arise and cause corrosion.

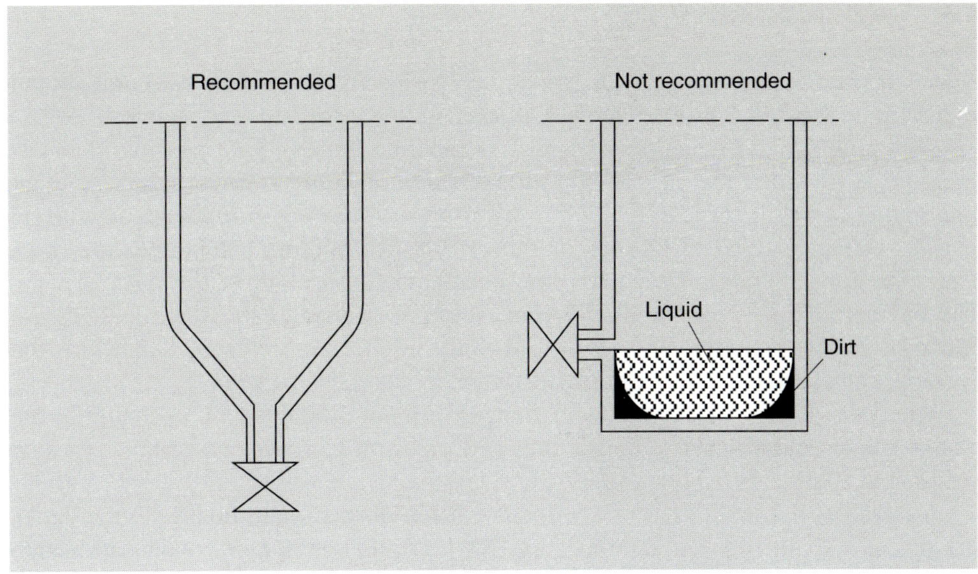

Fig. 101 Liquid containers should be designed such that they can, without unreasonable effort, be completely emptied. Dirt-collecting corners should be avoided.

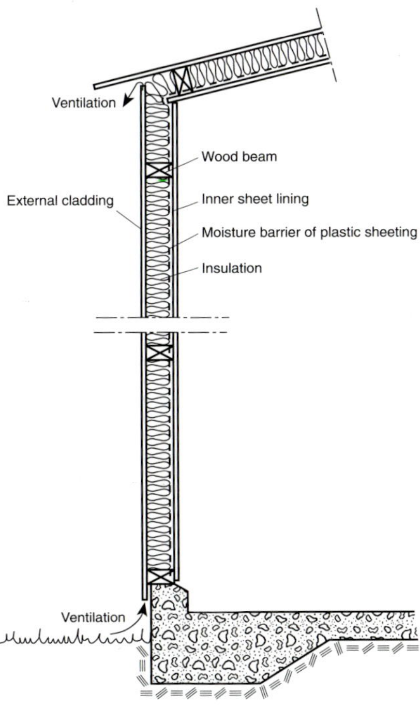

Fig. 102 Cross-section through an insulated external wall with outer and inner metal sheet coverings; the space inside the wall is well ventilated and a moisture barrier has been applied at the inner covering.

7.3 WATER IN CREVICES

Water in crevices can give rise to crevice corrosion in metals (see 4.3) and should therefore be avoided. This is especially important in constructions of stainless steel.

This problem can arise, for example, in pipeline flanges. One possible countermeasure is to widen the crevice so that exchange of liquid is encouraged. Another alternative is to fill the crevice, e.g. with a plastic jointing compound or with an elastic sealing material of plastics or rubber. Plastics jointing compounds are often based on polysulphides. They are sometimes applied on cloth in the form of tape. Elastic sealing strips can be made from neoprene-rubber. If the sealing strips are made of porous material, it is essential that the pores are closed. Otherwise the material can accelerate crevice corrosion when moisture is absorbed.

The risk for crevice corrosion in flanges can be eliminated by exchanging the flange for a welded joint although crevice corrosion can arise in certain types of welds (Fig. 103).

Another well-known case of crevice corrosion occurs when liquid storage tanks are placed on concrete foundations (Fig. 104). Liquid spillage or condensate which runs down the outer walls of the tank is carried by capillary action into the crevices between the foundation and the bottom of the tank, and can lead to crevice corrosion on the outside of the tank bottom. In this case crevice corrosion can be avoided by designing the foundation as shown in the figure. It is of further advantage to use a welded plate to lead away the liquid which runs down the outside of the tank (Fig. 104).

7.4 METALLIC CONTACT BETWEEN DISSIMILAR METALS IN MOIST ENVIRONMENTS

It is an old rule to avoid connecting metals of differing degrees of nobility, in order to avoid *bimetallic corrosion* (see 4.12). Care has, however, in many cases been exaggerated. Bimetallic corrosion can take place only if the points of contact are located in a moist environment. Risk of corrosion does not occur, for example, in a dry indoor atmosphere. The area ratio between the two metals is also of importance (see Fig. 49). As material for small fasteners one ought to choose a material having the same or a higher level of nobility than the surrounding metal. In a similar way, when welding, a filler metal having at least as high a level of nobility as the substrate material should be chosen.

When there is a risk of bimetallic corrosion, countermeasures should be taken, e.g. the introduction of an insulating material or painting the contact surfaces or the whole contact area and its surroundings (see 4.12).

Bimetallic corrosion can even occur as a consequence of water containing a more noble metal coming into contact with a less noble metal. A facade should therefore not be designed so that rainwater from a copper roof runs across a surface of aluminium or galvanised steel — or even bare mild steel.

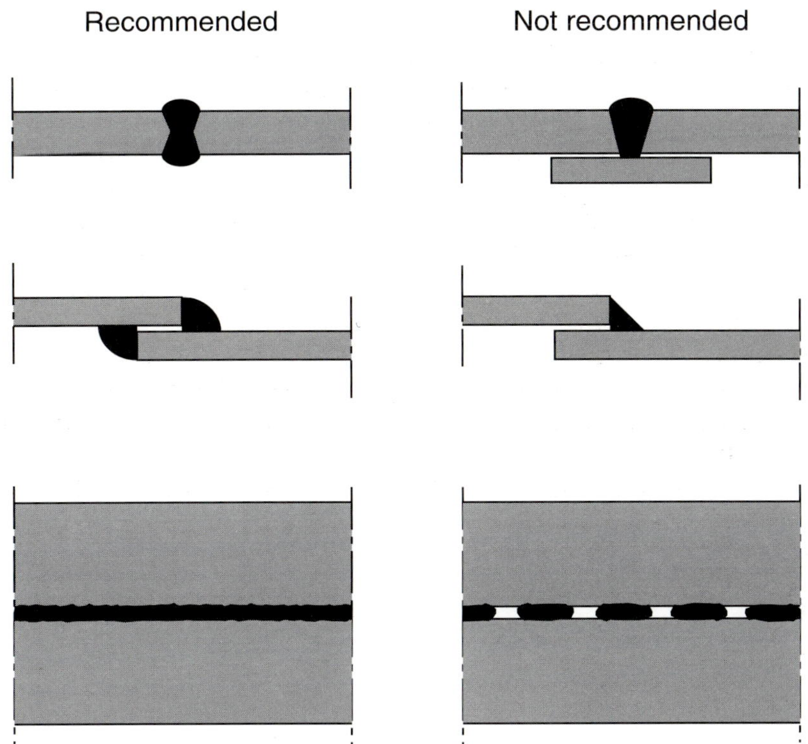

Fig. 103 Crevices should be avoided at welds, where moisture can accumulate and give rise to crevice corrosion.

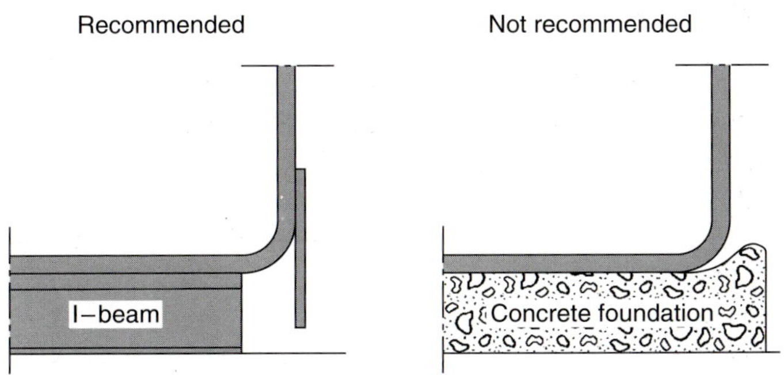

Fig. 104 Foundations for tanks should be designed so that the risk of crevice corrosion is avoided.

7.5 'STREAMLINE SHAPE' OF CONSTRUCTIONS WHICH ARE EXPOSED TO FLOWING LIQUIDS

When the surface of a metal is exposed to a corrosive liquid having too high a flow velocity, *erosion corrosion* can occur (see 4.8). The risk of erosion corrosion damage is especially great in places where there is a locally high flow velocity and considerable turbulence. This can be avoided by '*streamlining*' as far as possible in pipelines, heat exchangers and other constructions exposed to a flowing liquid. A branch joint on a pipe should thus be designed so that edges do not protrude into the stream and disturb the flow (Fig. 105). Further, when bending pipes, corrugations should be

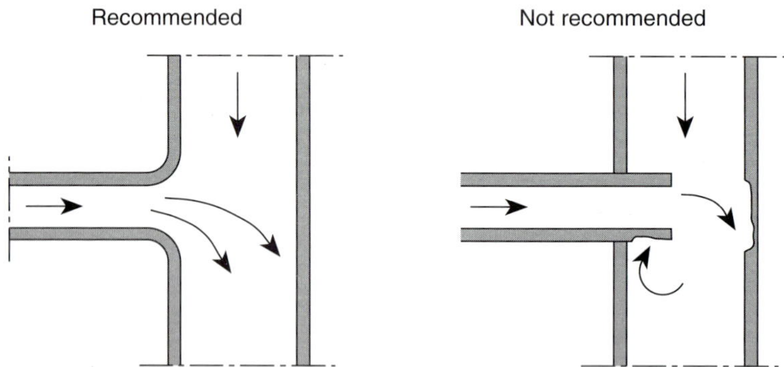

Fig. 105 At branch joints to pipes the edges ought not protrude into the stream and disturb the flow, since the resulting turbulence can cause erosion corrosion.

avoided, since damaging turbulence can arise at the corrugations (Fig. 106). Needless to say, dented pipes should be avoided. One should also take into account the risk of erosion corrosion from the flow of liquid into a container. The inlet pipe should be placed so that the incoming liquid stream is not excessively concentrated on to a restricted area of the container wall. To counteract such damage an easily replaceable baffle can be positioned in front of the inlet or the container wall can be strengthened at this point. There is also a risk of locally high flow velocities and consequent erosion corrosion effects even at regions of outflow from vessels.

7.6 'THERMAL BRIDGES'

The walls of houses with metal cladding often have a steel framework. If, for example, a steel member is in metallic contact with an outer metal cladding then the steel member can, because of its good heat conduction, act as a *thermal bridge*. When the outdoor temperature is low there is then a risk of water condensing on the inner parts of the structure. The condensed water so formed can lead to water damage

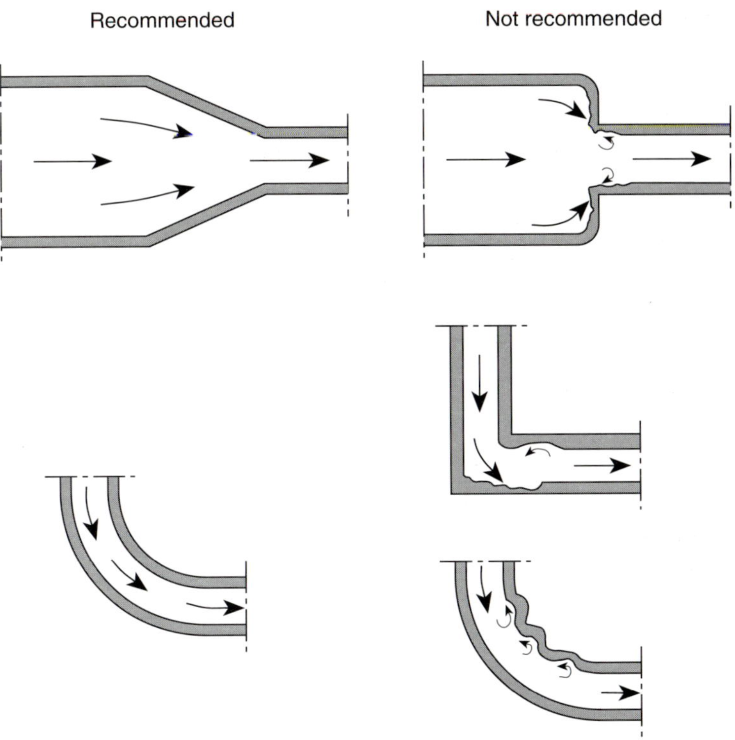

Fig. 106 'Streamlining' favours the avoidance of erosion corrosion.

including corrosion of the inner parts of the wall. Nevertheless, these 'thermal bridges' can be broken, and the trouble avoided by using a heat-insulating material between the outer cladding and the frame members (Fig. 107).

In a similar way condensation and corrosion can occur in gas-filled tanks built on heat-conducting supports. In this case too, the 'thermal bridge' can be broken by heat-insulation (Fig. 108).

7.7 DESIGN FOR SURFACE TREATMENT

When designing a construction which is to be protected from corrosion by surface treatment, it is important to avoid:

— non-draining hollows, narrow crevices, etc. where residues from the surface treatment bath can be held since these can later give rise to corrosion or even explosion, should the construction be subsequently coated by dipping in molten metal, and

— sharp edges and corners, deep recesses as well as sharp geometrical changes; otherwise there is a risk for unevenness in the surface coating at these sites.

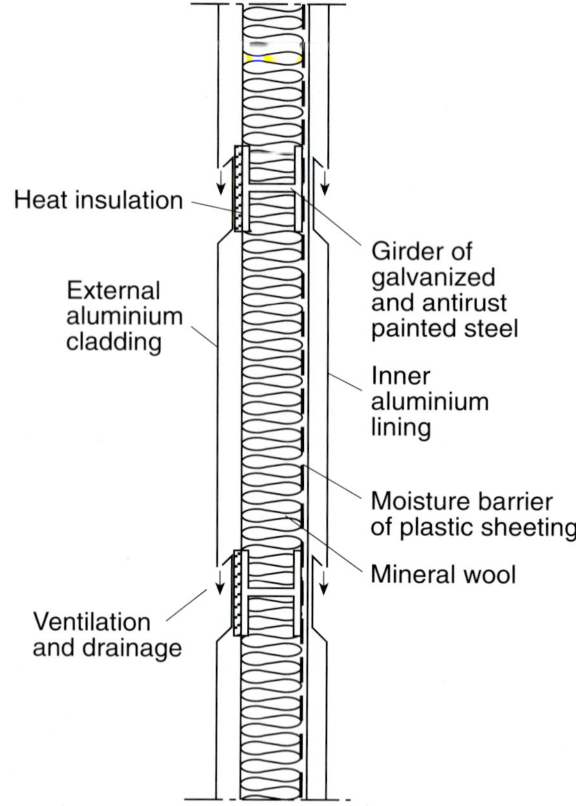

Fig. 107 Cross-section through a wall having inner and outer coverings of metal sheet; measures have been taken for drainage and ventilation of the inner space of the wall as well as for avoiding 'thermal bridges'.

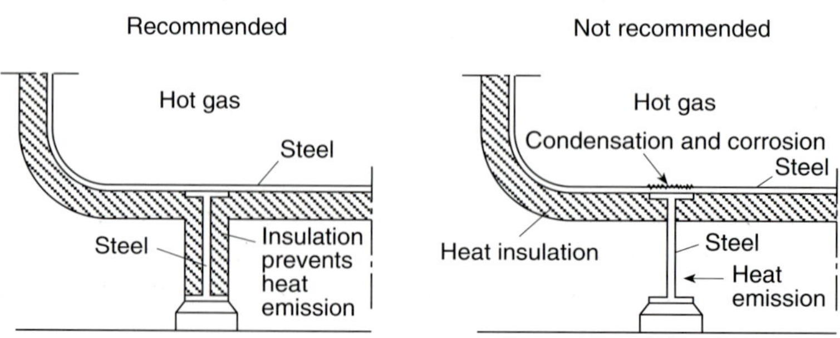

Fig. 108 Heat-conducting supports can cause condensation and corrosion in tanks containing hot gas.

8
The corrosion characteristics of the most common metals in use

Of all the metals in common use those that are ferrous-based, such as carbon steel, cast iron and stainless steel, occur most frequently. Non-ferrous metals such as aluminium and copper and their alloys are, however, also widely used as construction materials.

8.1 STEEL AND CAST IRON

8.1.1 Materials

Pure iron has only a minor technical usage. On the other hand, iron alloyed with carbon is the most used metallic constructional material. In *steel* (or carbon steel) the carbon content is up to about 1.3%, while in *cast iron* it is between 2 and 4%.

Micro-alloyed steels and carbon-manganese-steels are usually included in the category 'carbon steels'. Carbon-manganese steels often have a manganese content of up to 1.5%; the dividing line with low-alloyed steel being 'diffuse'. But when the total content of alloying substance (except for carbon) exceeds 5%, then the steel is referred to as high-alloyed steel.

Amongst the different variants of cast-iron one distinguishes between *grey* and *white cast iron* (Fig. 109). Grey cast iron is characterised by a phase of graphite, which can take the shape of flakes or spheres (spheroidal or ductile cast iron). White cast iron is characterised by a phase of cementite (Fe_3C). Malleable iron solidifies first to white cast iron and is then heat-treated so that irregular grains of graphite are formed. Cast iron alloyed with, for example, nickel or silicon (silicon iron), is also produced.

Because of their low price and high strength, steel and cast iron are widely used for structures in different environments:

— in the atmosphere, for example, buildings, bridges, pylons and cars,
— in water, for example, ships and offshore platforms, and
— underground, for example, water and gas pipes as well as petroleum and oil tanks.

8.1.2 General corrosion characteristics

Corrosion of steel and cast iron is called *rusting* and the corrosion products are called *rust.*

The potential-pH diagram for the system $Fe–H_2O$ (Fig. 110) provides an overview

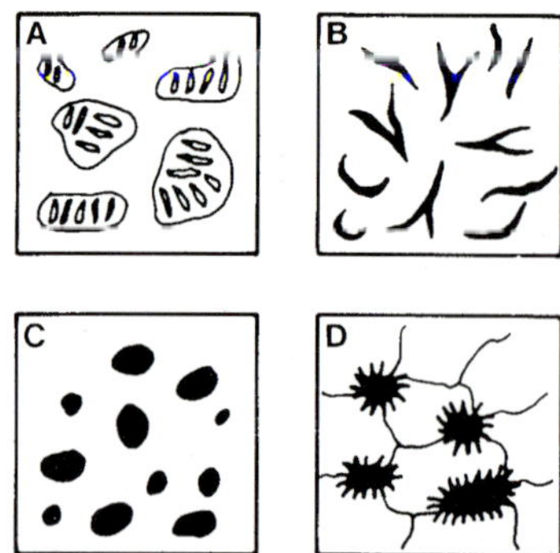

Fig. 109 Structures of different types of cast iron: A, white cast iron; matrix of cementite (Fe_3C) with precipitates of pearlite; B, grey cast iron; flakes of graphite in a matrix of pearlite or ferrite; C, spheroidal cast iron; nodules of graphite in a matrix of pearlite or ferrite; D, malleable iron; irregular grains of graphite in a matrix of pearlite or ferrite.

of the corrosion possibilities of steel and cast-iron under different environmental conditions. It can be seen that the metal is not immune in aqueous solutions. The domains of stability for Fe_3O_4 and F_2O_3 indicate the possibility of passivation at relatively high pH values (8-11). At extremely high pH values however, there is the risk of corrosion. Low concentrations of alloying additives do not generally have any significant effect on the corrosion characteristics.

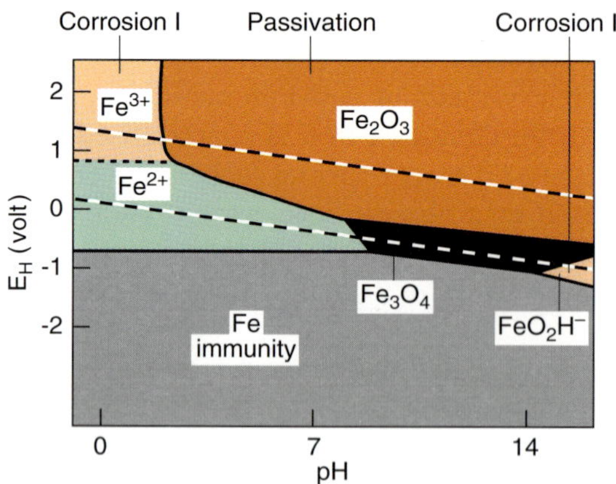

Fig. 110 — Potential-pH diagram for Fe-H_2O at 25°C; 10^{-6} M dissolved Fe [2].

8.1.3 Atmospheric corrosion

Steel rusts in an *outdoor atmosphere*. The coating of rust which is thus formed consists of a porous, loosely adhering, outer layer of crystalline α- and γ-FeOOH and a dense, more closely adhering inner layer of amorphous FeOOH and crystalline

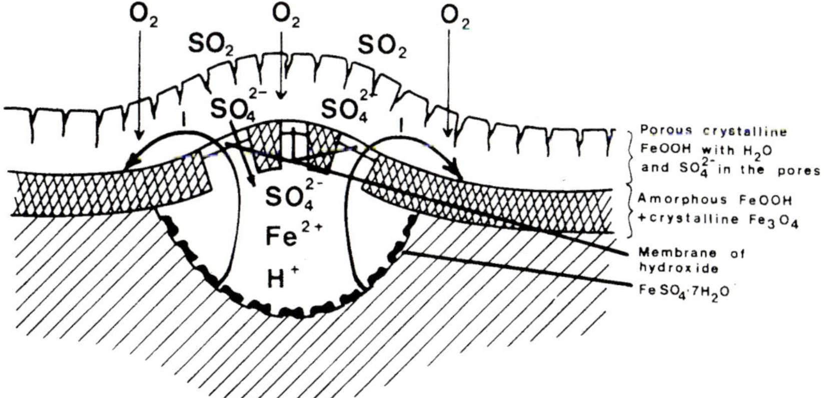

Fig. 111 Corrosion cell, 'sulphate nest', active on atmospheric corrosion of steel.

Fe_3O_4. This rust coating does not give good protection against further attack. Corrosion takes place by the action of corrosion cells, where the anodes are situated in *sulphate nests*, i.e. small pits rich in sulphate. The surrounding area acts as a cathode (Fig. 111). The reactions taking place can be summarised by the following formulae:

at the cathode

$$\tfrac{1}{2}O_2+H_2O+2e^-\rightarrow 2OH^- \quad \text{and}$$

$$Fe^{3+}+e^-\rightarrow Fe^{2+},$$

at the anode

$$Fe\rightarrow Fe^{2+}+2e^-, \quad \text{and}$$

in the rust

$$2Fe^{2+}+3H_2O+\tfrac{1}{2}O_2\rightarrow 2FeOOH+4H^+ \quad .$$

The increase in depth of penetration (P) with time (t) generally follows a power law:

$$P=kt^n ,$$

where k and n are constants. This equation can also be written

$$\log P=\log k+n \log t.$$

The time functions of pit depth in three different types of atmosphere are shown in Fig. 112. The same curves are represented by straight lines in a bilogarithmic diagram (Fig. 113). At steady state conditions the corrosion rate for steel in different

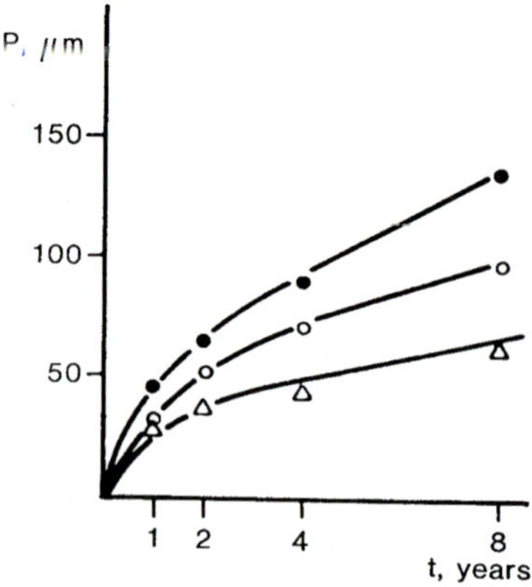

Fig. 112 Corrosion depth (P) as a function of exposure time (t) on atmospheric corrosion of steel: ●, Mülheim, Ruhr; ○, Cuxhaven; △, Olpe [15].

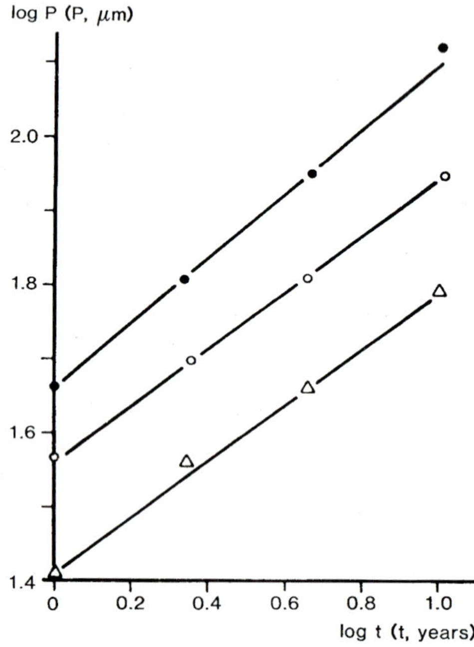

Fig. 113 Bilogarithmic diagram of corrosion depth (P) and exposure time (t); shows the same results as Fig. 112 [15].

types of atomosphere are:

rural atmosphere	5–10 μm/year
marine atmosphere	10–30 μm/year
urban atmosphere	10–30 μm/year
industrial atmosphere	30–60 μm/year

Steel does not usually rust *indoors* if the relative humidity is less that 60% (see 5.3). A special type of serious atmosphere corrosion can, however, occur indoors after a PVC fire. When PVC (polyvinylchloride) burns hydrogen chloride (HCl) is formed, which is gaseous and readily spreads to adjacent areas. In presence of water vapour this will form hydrochloric acid which will rapidly attack metal surfaces, e.g. nail-heads in walls, machines, electrical equipment or stored goods. Therfore, after a PVC fire one should immediately survey the spread of hydrochloric acid on the premises and carry out cleaning and protection measures where pollution exceeds a certain limit, which for steel is usually 10μg Cl^- per cm^2 and with more robust structures, 20μg Cl^- per cm^2.

8.1.4 Corrosion in water

The corrosion of steel in water is mainly controlled by the cathode reaction, i.e. usually the supply of oxygen, but the pH value of the water and its ability to precipitate protective calcium carbonate are also important (see 5.1). In closed heating systems for example, where the content of dissolved oxygen in the water is soon used up by corrosion, the corrosion rate is negligible. In sea or fresh water with a high oxygen content, uniform corrosion takes place often at the rate of 50–150 μm/year. In addition, localised corrosion can occur at a considerably higher rate, for example, in the splash zone at sea-level, underneath fouling organisms, at crevices, or where the flow velocity of the water is high. Microorganisms can also accelerate corrosion in steel, even under anaerobic conditions.

Cast iron structures are often less sensitive to corrosion than steel constructions. This is partly dependent on their usually heavier dimensions and a greater corrosion allowance. Cast iron can, however, be attacked by graphitic corrosion (see 8.1.7).

8.1.5 Corrosion by the action of differential aeration cells

The major part of the corrosion damage which occurs on the bodies of motor vehicles starts from the inside and is mostly localised to places where road mud accumulates (see Fig. 28). Under the mud deposit the supply of oxygen is reduced. The surface there acts as the anode and is attacked, while areas with better oxygen supply act as the cathode in the corrosion cell. Such corrosion damage has been and is still largely responsible for the depreciation in value of vehicles and is among the corrosion effects of greatest economic importance in the community. Fortunately an improved awareness of the causes of this problem and the use of good corrosion prevention techniques mean that this problem is perhaps not as serious as it was in the 1960s and 1970s.

Underground structures made of steel or cast iron often suffer from pitting by the action of differential aeration cells. This is because the surface is exposed to soil conditions which vary in their oxygen permeability (see 5.2).

Corrosion by the action of differential aeration cells also occurs in constructions which are heat insulated with porous material, e.g. mineral wool or polyurethane foam, if they are exposed to water. Such damage occurs in culverts in district heating networks. These consist of a central steel pipe surrounded by an insulating material, which in turn is surrounded by a protective cover of, typically, concrete or plastics. If water gets into the insulation, via leaky joints in the protective cover, or by some other route, then differential aeration cells can be set up and attack of the central steel pipe will result. A similar type of corrosion can occur in heating pipes in buildings when the insulation has become wet, e.g. through rainfall during construction or through leakage from a joint (see Fig. 29). In some cases the heating pipes have been perforated even before the building has been completed.

8.1.6 Stress corrosion cracking

Carbon steel can be attacked by *stress corrosion cracking* in alkaline solutions at high temperatures. This kind of attack has occurred, for example, in continuous kraft digesters where the steel is exposed to white liquor containing sodium sulphide and sodium hydroxide at a temperature of 150–170°C.

Underground steel piping can also be damaged by stress corrosion cracking. This has occurred in piping at raised temperatures, e.g. in district heating networks and in those parts of high-pressure piping for gas which are situated next to the compressor stations. In the latter case the resulting tensile stresses were caused by the gas pressure in the pipes. The stress corrosion cracking is made possible by the presence of $NaHCO_3$, OH^- or NO_3^- in the soil. Cathodic protection can also have an influence, partly because the electrode potential of the pipe is regulated to a level which is conducive to cracking, and partly because OH^- ions are formed at the surface of the cathode.

Even ammonia can give rise to stress corrosion cracking in steel and cracking has occurred in steel tanks used for liquid ammonia.

8.1.7 Graphitic corrosion of grey cast-iron

Graphitic corrosion takes place in, for example, pipes and valves of cast iron, which are exposed to sulphate-rich soil, sea water, mine water etc. In the corrosion process it is the iron that is attacked, leaving behind a residue of graphite and corrosion products (Fig. 114). The object can keep its original shape in spite of the attack but both strength and mass are considerably reduced. Cast-iron pipes which have suffered from graphitic corrosion can therefore be brittle and leaky (see Fig. 2).

8.1.8 The corrosion of steel in concrete

Steel embedded in concrete is well protected from corrosion provided that the concrete has a sufficiently low porosity, that it is free from cracks and that the layer of concrete over the steel is sufficiently thick. Usually a thickness of 20 mm is recommended. The steel surface in contact with the alkaline concrete acquires a passivating layer. The protective layer, however, can be damaged if the concrete is carbonated through reaction with carbon dioxide from the environment leading to a decrease of the pH

Fig. 114 Graphitic corrosion (dark area) of the outside of a water pipe of grey cast-iron after 21 years in clay with a resistivity of 1300 ohm cm (×1/1) .

value. Chloride originating from e.g. sea water, or added during casting may also be detrimental. If the steel surface becomes exposed at some place, for example at a crack in the concrete or at non-embedded parts, an activated steel surface can appear, which together with the passivated steel surface in the concrete, can form a so-called *active–passive cell* (Fig. 115). In this case the activated surface becomes the anode and is attacked locally, and the passivated steel surface becomes the cathode. The action of such an active/passive cell sometimes leads to perforation of water pipes embedded in concrete. Other examples include the corrosion of the steel reinforcements in concrete balconies, concrete bridges and other concrete structures.

In order for the concrete to have a sufficiently low porosity it is recommended that the water content of the cement mixture is limited. The water/cement ratio of the concrete should in general not exceed 0.6 and when embedding piping it should even

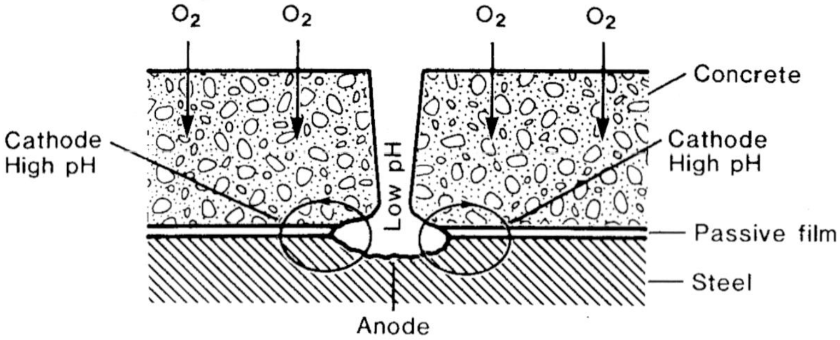

Fig. 115 Active/passive cell on steel embedded in concrete.

be kept below 0.5. Good compaction is also desirable and can be achieved by vibration of the mix. Wide cracks in outdoor environments should be taken care of by injection. Cracks with a width below 0.1 mm are, however, usually judged to be safe.

8.1.9 Corrosion protection

Under corrosive conditions the corrosion rate of steel and cast iron is relatively high (the rust formed can also stain nearby surfaces). In certain cases the low corrosion resistance can be compensated for by over-dimensioning—the so-called *rust allowance*. But usually some form of corrosion prevention is preferred, e.g.:

— anti-rust painting,
— coating with plastic, as with sheet metal for building purposes,
— coating with metal, such as zinc, aluminium, aluminium-zinc or nickel,
— dry-air storage,
— addition of corrosion inhibitor to the corrosive environment, or
— cathodic protection of constructions in aqueous environments.

These countermeasures are described in separate sections.

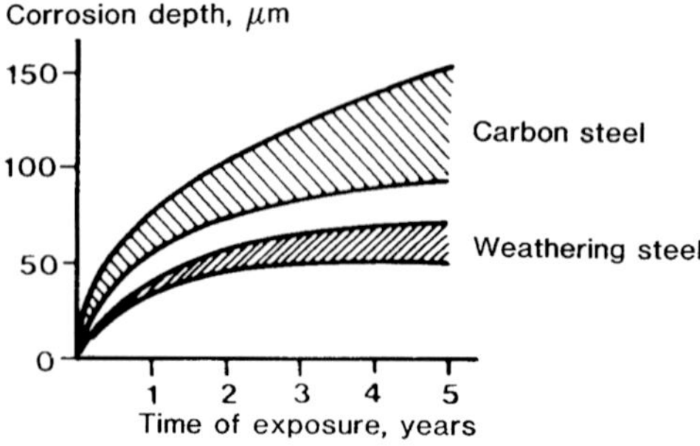

Fig. 116 Corrosion depth as a function of exposure time on weathering steel and carbon steel in an atmosphere having a SO_2 deposition rate of about 90 mg dm^{-2}/day [16].

Fig. 117 Chimney of weathering steel at a district heating plant.

8.1.10 Weathering steel

The resistance of steel to atmospheric corrosion can be somewhat improved by low alloying additions of elements such as chromium, phosphorus and copper to produce the so-called *weathering steels*. These usually contain 0.25–0.5% Cu, 0.04–0.15% P, 0.2–0.9% Si, 0.3–1.2% Cr and up to 0.6% Ni. On exposure to the atmosphere the rust on weathering steels gives a degree of protection, provided the conditions are favourable (Fig. 116). The rust then 'matures' slowly over the years to a decorative, *blue-brown patina* and anti-rust painting becomes unnecessary. But on surfaces which are continuously wet, exposed to a marine environment or not washed by rain, i.e. in sheltered conditions in which aggressive salts can accumulate on the surface of the metal, the protective patina is generally not formed. Weathering steel has been found to be suitable for constructions which are alternately exposed to moist and dry conditions, e.g. lattice trussed structures for switchgears and outer mantles to chimneys (Fig. 117).

8.2 STAINLESS STEEL

8.2.1 Materials

Stainless steels are iron-based alloys in which chromium is the main alloying additive and at a concentration of at least 12%. Because of the chromium content, stainless steels are easily passivated and hence have good corrosion resistance in many of the more common environments. In unfavourable conditions, however, even stainless steels can be subject to, for example, uniform corrosion, pitting, crevice corrosion, intergranular corrosion or stress corrosion cracking.

One can distinguish between different types of stainless steels depending on their structure, e.g. ferritic, austenitic and ferritic-austenitic steels. Differences in structure convey differences in corrosion characteristics and even differences in weldability, hardening capacity and magnetic properties. Ferritic and ferritic-austenitic steels are magnetic in contrast to austenitic steels. Table 11 provides a survey of some stainless steels of interest from the corrosion point of view, and their corrosion characteristics.

8.2.2 The active and passive states

The conditions for passivation are shown in the anodic polarisation curves of the steels (Fig. 118). If, in the case of a stainless steel in sulphuric acid solution, the electrode potential is increased then the current density rises to a maximum with dissolution of the metal taking place in the *active state*, the current density being an expression of the dissolution rate. At a certain potential, the *passivation potential*, the corrosion current density is drastically reduced and the metal surface becomes *passivated*. Passivation is associated with the formation of a thin, protective coating which largely consists of a mixed iron-chromium oxide and hydroxide. If the potential is further increased to very high values the current density will increase again as a result of, so-called, *transpassive corrosion*. Corrosion in the active as well as the transpassive states is usually uniform. The resistance to uniform corrosion for different stainless steels in a specific chemical agent is often shown in an iso-corrosion diagram where the corrosion rate of 0.1 mm/year is indicated as a function of solution concentration and temperature (Fig. 119).

8.2.3 Pitting and crevice corrosion

When using stainless steel in an environment containing a high chloride content, such as sea water or the bleaching liquors used in the pulp industry, *localised corrosion* will often occur in the form of pitting (Fig. 120) — which can sometimes result in perforation of pipe walls — or in the form of *crevice corrosion*, e.g. in flanged joints (see Fig. 26). These two corrosion types are related.

Stainless steels are susceptible to localised corrosion, especially in the presence of halogen ions. Amongst these the chloride ion is the most corrosive and also of greatest practical importance.

Table 11 Survey of some stainless steels often used under corrosive conditions; the higher the number, the better the resistance

Designation Sweden	Nominal composition, % (bal.Fe)	Equivalence		Resistance			Application examples
		EN 10088	ASTM UNS	Pitting and crevice corrosion in Cl^--containing environment	Intergranular corrosion after welding	Stress corrosion cracking in Cl^--containing environment	
Ferritic							
SS 2320	17Cr, max 0.10 C	1.4016	S 43000	1	1	4	Kitchenware
SS 2326	18Cr, 2.2Mo, max 0.025C, Ti	1.4521	S 44400	2	4	4	Water heaters
Austenitic							
SS 2333	18Cr, 9Ni, max 0.05 C	—	—	1	3	1	Equipment in chem. ind.
SS 2343	17Cr, 12Ni, 2.7Mo, max 0.05 C	1.4436	—	2	3	1	Equipment for corrosive media
SS 2562	20Cr, 25Ni, 4.5Mo, 1.6Cu, max 0.025 C	1.4539	N 08904	3	4	3	Equipment for very corrosive media
SS 2378 Avesta 254 SMO	20Cr, 18Ni, 6.2Mo, 0.7Cu, 0.2N, max 0.020 C	1.4547	S 31254	4	4	3	Bleaching equipment in pulp industry
Avesta 654 SMO	24Cr, 22Ni, 7.3Mo, 0.5Cu, 3.5Mn, 0.5N, max 0.015C	—	S 32654	4	4	3	In chem. ind. with chloride and strong acid
SS 2584 Sandvik Sanicro 28	27Cr, 31Ni, 3.5Mo, 1.0Cu, max 0.020 C	1.4563	N08028	3	4	4	Production of fertilisers
Ferritic-austenitic							
SS 2324	25Cr, 5Ni, 1.5Mo, max 0.10 C	1.4460	S 32950	2	2	3	Equipment in chem. ind.
SS 2327	23Cr, 4Ni, max 0.03 C	1.4362	S 32304	2	3	4	In chem. ind.
SS 2377	22Cr, 5.5Ni, 3.0Mo, 0.15N, max 0.030 C	1.4462	S 31803	3	4	3	Heat exchangers for gas and oil

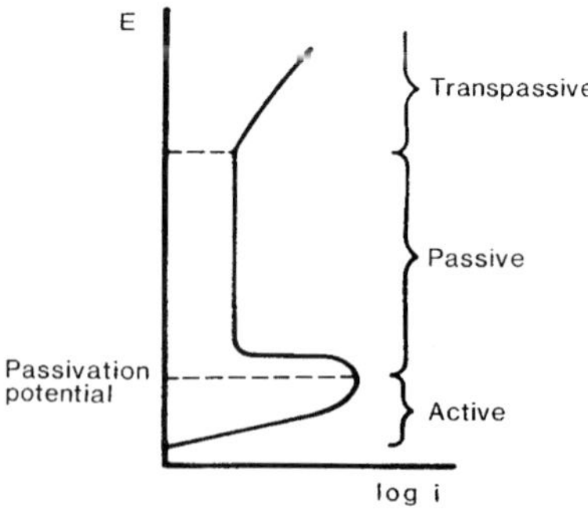

Fig. 118 Anodic polarisation curve for stainless steel in a sulphuric acid solution.

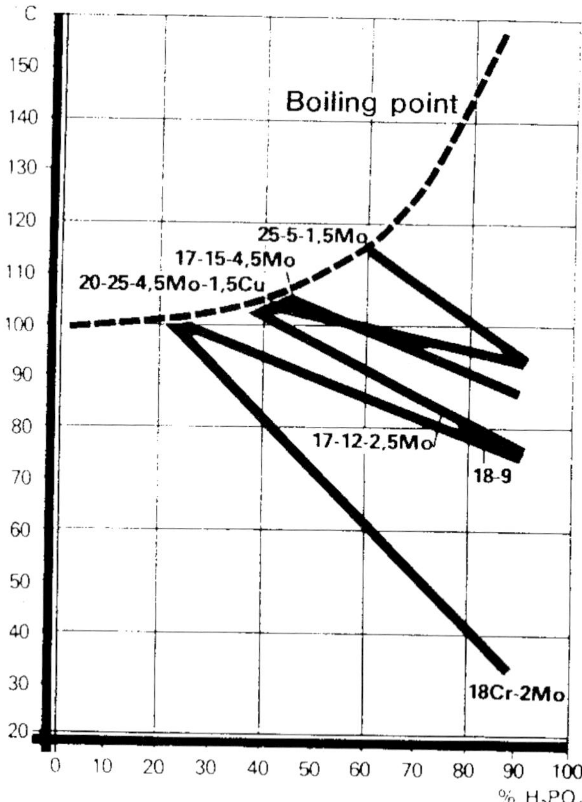

Fig. 119 Iso-corrosion curves for stainless steels in phosphoric acid; ——— gives the conditions for a corrosion rate of 0.1 mm/year (below the line the corrosion rate is lower); – – – gives the boiling point of the acid solution [17].

Fig. 120 Pits in a stainless steel pipe (18Cr, 9Ni).

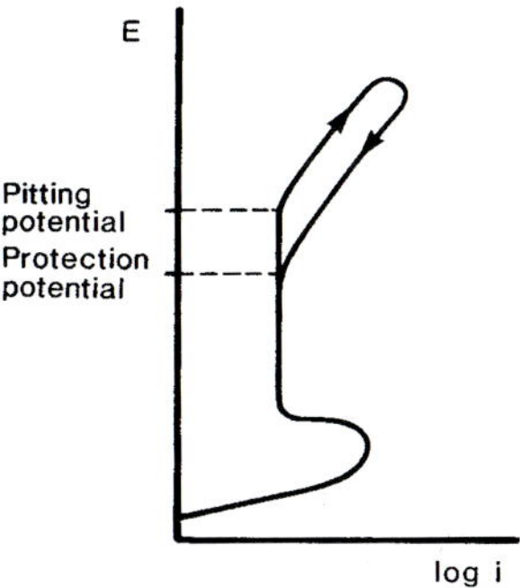

Fig. 121 Anodic polarisation curve for stainless steel in sulphuric acid solution containing chloride.

Addition of sulphide considerably increases the corrosivity of the environment and may take place, e.g. due to the dissolution of sulphide inclusions in the steel surface.

Pitting in stainless steel also affects the shape of the anodic polarisation curve (Fig. 121). Thus, if the potential is increased above a certain critical value, referred to as the breakdown potential, the current density will begin to increase and the curve often shows thereafter a series of current peaks. Since this rise marks the beginning of pitting the breakdown potential is in this case called the *pitting potential.* If the potential is then decreased, passivation is achieved again, but only when a *protection potential,* which is a little below the pitting potential, is reached. A similar development occurs with corrosion in crevices or under surface deposits. The existence and value of the pitting potential can be demonstrated by using an auxiliary electrode and an applied voltage. In practice, the presence of an oxidising agent, e.g., oxygen, chlorine or peroxide in the solution is often sufficient to raise the potential to the pitting value with consequent attack. The breakdown potential is not a well-defined constant but depends to a large extent on conditions such as chloride concentration, temperature and the method of measurement.

Localised corrosion is observed only after a certain incubation time during which the initiation of the attack takes place. This is followed by the propagation stage and the growth of the pit. Both the initiation and the propagation take place by a mechanism which involves electrochemical corrosion cells.

Many mechanisms have been proposed for the *initiation stage*. In some the initiation can consist of a new, non-passivated metal surface being created as a result of the dissolution of sulphide or sulphide/oxide inclusions in the surface. This could then lead to an acidic solution containing a high sulphide content developing in the incipient cavities; these conditions would not allow passivation of the stainless steel surface to occur at these sites. In other cases, initiation can be associated with depletion of oxygen in a crevice or under a deposit. This would give rise to an oxygen concentration cell with the anode in the crevice or under the deposit, and the cathode outside these regions. Hydrolysis of anodically dissolved metal ions in the various situations gives rise to acidic, non-passivating conditions at anodes.

When localised attack has been initiated and the growth has reached steady state conditions then the process can be said to have reached the *propagation stage*. Certain characteristic conditions now prevail in the pit (Fig. 122):

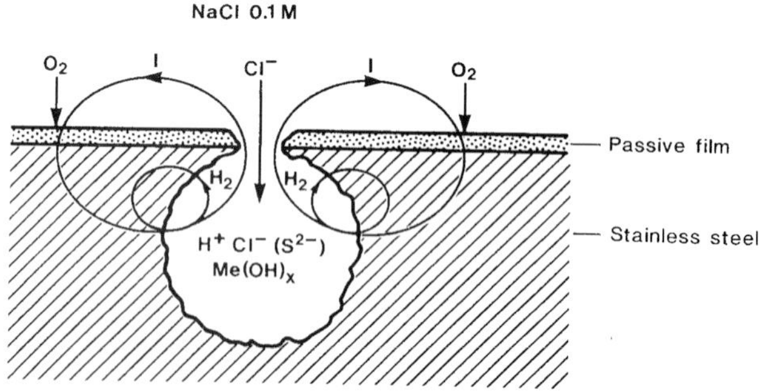

Fig. 122 Pitting in stainless steel.

— The pH value is lower than in the bulk of the solution because anodically dissolved metal ions, such as Fe^{2+} and Cr^{3+} have been hydrolysed with the

formation of oxides, hydroxides or hydroxide salts thus releasing hydrogen ions. The actual pH value will depend on the composition of the steel and will be lower the more corrosion resistant the steel. Often a pH value of 0–1 can arise.
— The solution in the pit has a higher chloride concentration than the bulk of the solution. This is because chloride ions migrate against the electric current in through the mouth of the pit. In practice a chloride concentration as high as 5 M can occur in the pit.

Even during the propagation stage the cathodic reaction mainly consists of the reduction of oxygen on the passivated area around the pit. Some reduction of hydrogen ions to hydrogen gas can also take place, especially in the pits, where the pH value is low. The anodic current density in the pits is several powers of ten higher than the cathodic current density on the outer surface. However, the growth of the pit can be stopped if the conditions in the pit are changed so that *re-passivation* commences.

There are several possibilities for *counteracting* localised corrosion in stainless steel:
— By using *cathodic protection* the electrode potential can be reduced and held at a value lower than the protection potential for the steel. The potential can be regulated by polarisation with the aid of an external source of current and an auxiliary electrode, by contact with a less noble material or by adjustment of the redox potential of the corrosive environment.
— By choosing a suitable stainless steel one can reduce the risk of localised corrosion. The resistance increases with the chromium content. The inclusion of molybdenum or nitrogen in the alloy also has a favourable effect.
— A reduction of the chloride concentration in the medium helps counteract localised corrosion.
— An increase in pH value also reduces the risk of localised corrosion.
— The addition of certain inhibitors is beneficial. Anodic inhibitors such as nitrite and chromate are, however, dangerous. If localised corrosion has already been initiated then these can accelerate the process. One should instead choose a cathodic inhibitor.
— Improved convection contributes towards a levelling-out of the concentration differences between the pits and the surroundings and therefore facilitates repassivation of the pits. At flow velocities above 1.5 ms^{-1} the risk of localised corrosion is very small but it can remain in narrow crevices, e.g. in flanged joints.
— The design of the construction should be such that crevices and deposits are avoided. Flanged joints could, for example, be replaced with welds; scale formed during welding should be removed and piping where mud is deposited or where organic growths occur should be cleaned regularly.

The resistance of stainless steels towards localised corrosion can be evaluated by:

— determination of the breakdown potential, i.e. the pitting potential or crevice corrosion potential; this can be done by recording the anodic polarisation curve in a solution of chloride (ASTM G 61),

— determination of the critical pitting temperature, (c.p.t.), and the critical crevice corrosion temperature, (c.c.t.); the c.p.t or c.c.t., is the lowest temperature at which attack takes place, while maintaining the stainless steel at a constant potential,

— total immersion testing in a suitable corrosive agent, e.g. ferric chloride solution (ASTM G 48–76), followed by measurement of the maximum pit depth; in a crevice corrosion test an elastic string or plastics disc is pressed against the test piece and the depth of attack is measured at the points of contact after the exposure period.

8.2.4 Intergranular corrosion

Intergranular corrosion can occur in certain types of stainless steel which have a high carbon content (0.05–0.15% C). It can take place if the stainless steel is heat-treated so that chromium carbide precipitates in the grain boundaries and the material is subsequently exposed to an acidic solution or sea water. The reaction mechanism is shown in Fig. 123. The precipitation of chromium carbide takes place only under

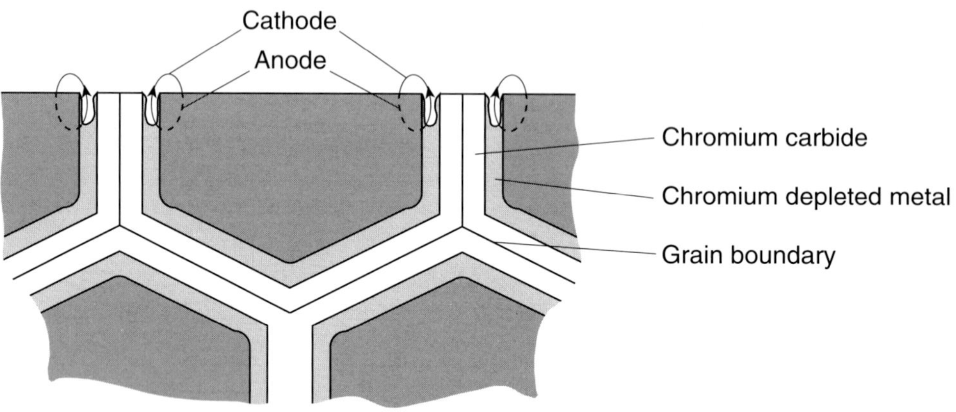

Fig. 123 Intergranular corrosion of stainless steel.

certain conditions, for austenitic steel this is in the temperature range 550-850°C. The steel is then said to be *sensitised*. The carbide precipitation results in a narrow zone near the grain boundaries being so depleted in chromium that the steel loses its 'stainless' character. Sensitisation is not restricted to manufacturing heat-treatment processes. It can also occur as a result of welding (see 8.2.5) (Fig. 124), since a region of the metal near the weld will pass through the above temperature range. On exposure to a corrosive environment the chromium-depleted zones, together with the

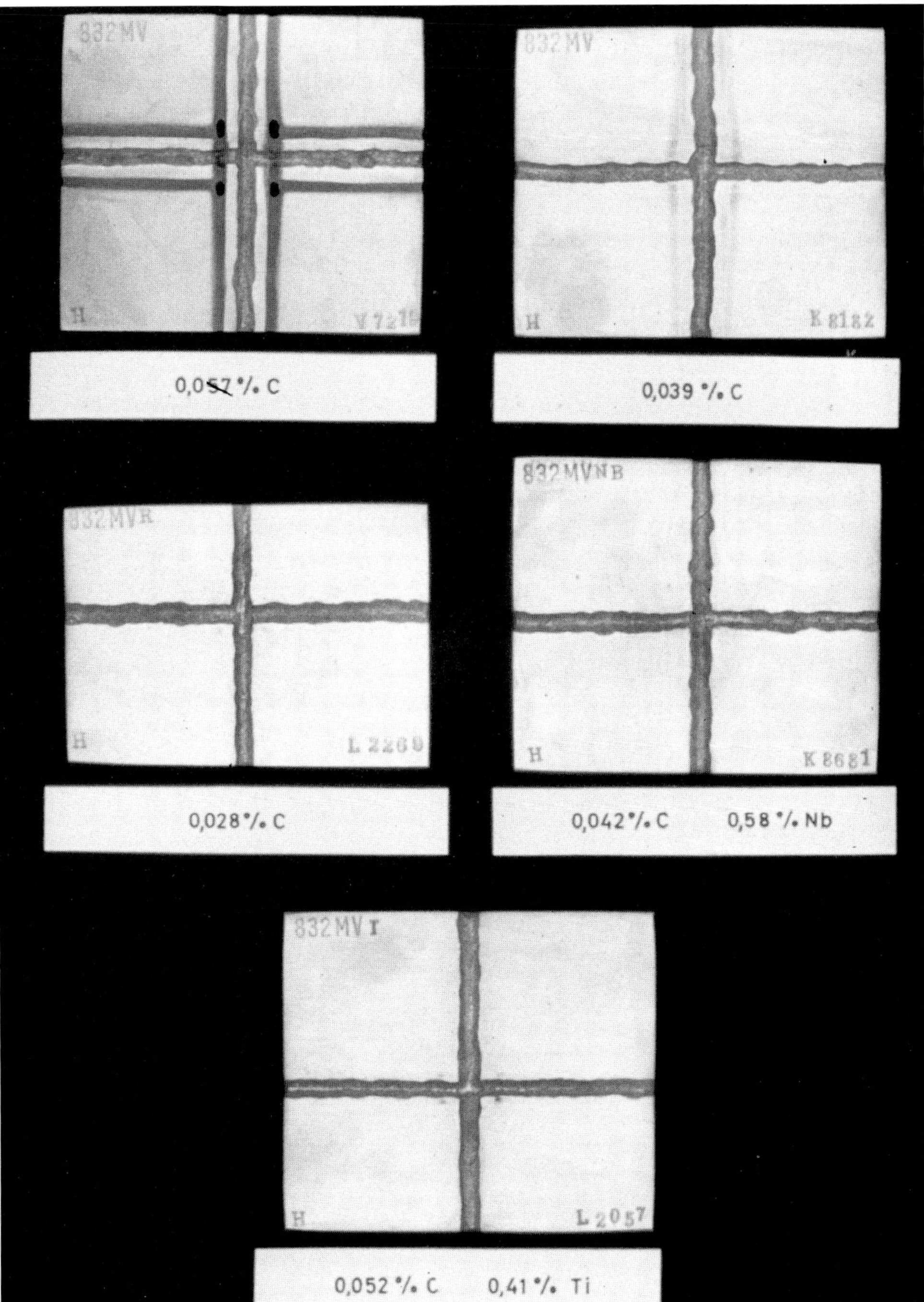

Fig. 124 Attack by intergranular corrosion along TIG welds in stainless steel sheet (18Cr,9Ni) after 4.5 hours' exposure to a solution with 15% HNO_3 and 3% HF at 80°C; the attack appears as a dark zone parallel to the welds; the effects of reducing the carbon content or adding Nb or Ti respectively are shown (Avesta AB).

remaining parts of the grains, form corrosion cells. In these the chromium-depleted metal acts as the anode and is attacked resulting in intergranular corrosion. Intergranular corrosion does not usually influence the shape of the object but the strength characteristics can be catastrophically reduced. In a chloride environment, e.g. sea water, the attack can be visible as pitting. Intergranular corrosion also results in the loss of metallic sound when the affected component is struck.

Countermeasures are designed so as to counteract the precipitation of chromium carbide, which is the cause of the sensitisation. The following possibilites exist:

— the choice of stainless steel with a low carbon content (<0.05% or preferably <0.03%),
— the choice of steel which has been *stabilised* by the addition of titanium or niobium which bind the carbon as carbides so that the formation of chromium carbide is prevented,
— the shortest possible time at 550-850°C (Fig. 125), and
— *solution heat-treatment*; the material is heated to about 1050°C to dissolve precipitates of chromium carbide formed, followed by subsequent rapid cooling so that new chromium carbide precipitates have no time to form.

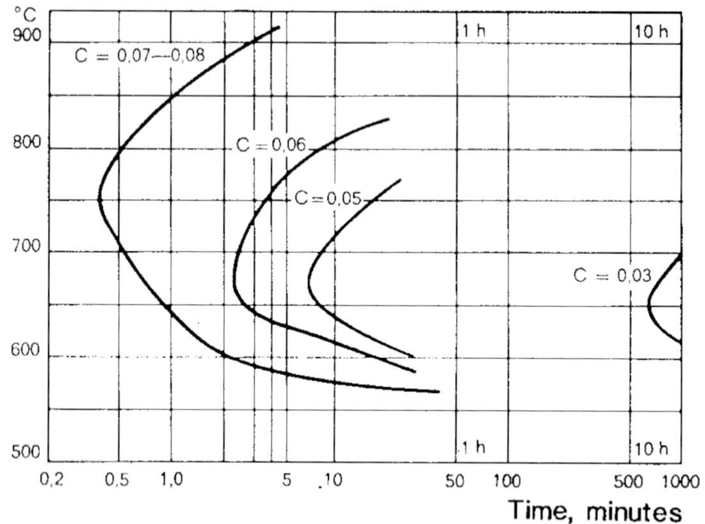

Fig. 125 Time-temperature diagram, showing the conditions for sensitising stainless steels (18Cr, 9Ni) having different carbon contents, under annealing conditions. The curves are based on corrosion testing in a boiling solution containing 10% H_2SO_4 and 10% $CuSO_4$ for 216 hours. Sensitisation takes place with conditions to the right of the curves (Avesta AB).

The resistance of stainless steel to intergranular corrosion can be evaluated through:

— *Strauss test* (SIS 11 71 05, ASTM A 262 Practice E); the test piece is exposed to a boiling copper sulphate solution containing sulphuric acid and metallic copper;

after exposure the test piece is deformed by bending, and the surface is then examined for cracks,

— *Streicher test* (ASTM A 262 Practice B); the test piece is exposed to a boiling ferric sulphate-sulphuric acid solution for 120 h; after the exposure the mass loss of the test piece is determined, and

— *Huey test* (ASTM A 262 Practice C); the test piece is exposed to a boiling 65% nitric acid solution; evaluation consists of determining mass loss; Huey testing is particularly significant for material that is to be used in strongly oxidising media.

8.2.5 Weld corrosion

After welding — even after assembly welding — the welds should be cleaned from oxides formed during heating and the surface zone — which is possibly chromium-depleted — should be removed; this is to create favourable prerequisites for passivation so that corrosion at the welds can be avoided. This cleaning can be achieved by pickling or grinding.

Even if the welds are cleaned they can in certain stainless steels and under unfavourable conditions have a lower corrosion resistance than the rest of the construction. Three main types of attack can be distinguished (Fig. 126):

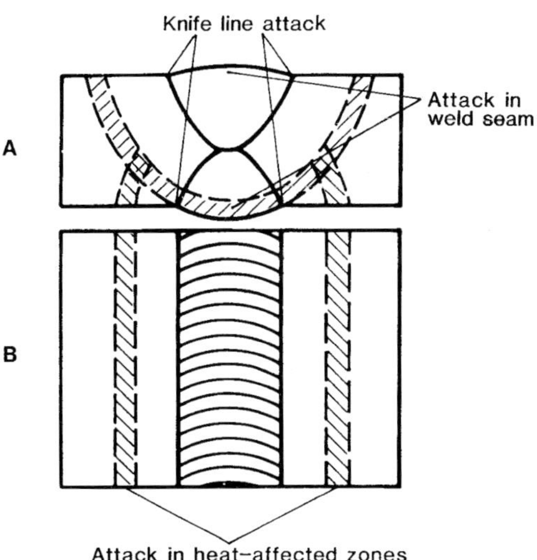

Fig.126 Localisation of different types of weld corrosion. A,cross-section; B, from above.

— attack on the weld seam, which consists of filler and substrate metal melted during welding,

— attack in heat-affected zones of the substrate due to welding, and

— knifeline attack; corrosion in a narrow zone of the substrate at the border between the weld seam and the substrate.

Attack on the weld seam can have the character of localised corrosion and arises when the filler metal is less noble than the substrate; this is a result of an unfavourable anode/cathode area ratio in the corrosion cell since the weld seam which constitutes the anode is small in area compared with the cathodic substrate. This problem can be avoided by choosing a filler metal which has at least the same degree of nobility as the substrate.

Attack in heat-affected zones. When welding, an area along the weld will be affected by heat which can cause sensitisation in certain stainless steels. This sensitivity can be eliminated by solution heat treatment after welding. Steels which are stabilised with titanium or niobium are not prone to sensitisation.

Knifeline attack occurs at the border between the weld and the substrate in stabilised stainless steels; this also applies to molybdenum-alloyed stainless steels. Several mechanisms can lead to this kind of attack. Knifeline attack has, however, lost some of its importance now that steels having an extra low carbon content have come more into use.

8.2.6 Stress corrosion cracking

A type of corrosion which can rapidly lead to serious damage is *stress corrosion cracking*. In order for mechanical tensions to initiate stress corrosion cracking they must exceed a critical level which will depend on several factors such as the composition of the stainless steel, the surface roughness, the grain size and structure as well as the environment and the temperature. Tensile stresses in a construction can arise from various causes, including welding and machining.

One can distinguish between two different types of stress corrosion cracking in stainless steel: intergranular and transgranular.

Intergranular stress corrosion cracking has received special attention in nuclear reactors of the BWR (Boiling Water Reactor) type. The corrosion environment in these consists of very pure water having a certain oxygen content (*ca* 0.2 ppm). Cracking has occurred especially at high temperatures (200–300°C). As a rule the tensile stresses have been high, above the yield strength. In most cases intergranular stress corrosion cracking has occurred in steels which were in the sensitised condition.

Transgranular stress corrosion cracking occurs mostly in environments having a high chloride concentration, but can also be found in the presence of concentrated alkalies. In practice, stress corrosion cracking is sometimes found in association with locally increased chloride concentrations on hot surfaces as a result of evaporation (Fig. 127). A high temperature is essential for transgranular stress corrosion cracking; thus, the chloride variant seldom occurs below 60°C and the alkaline variant seldom below 100°C. Transgranular stress corrosion cracking can occur in austenitic stainless steels but the resistance increases with the nickel content, so that alloys having 40% or more nickel can be regarded as practically safe. Ferritic chromium steels having low contents of other alloying constituents are generally regarded as being resistant to this kind of attack. Steels having the ferritic-austenitic structure are generally less sensitive than purely austenitic steels.

The mechanisms of stress corrosion cracking in stainless steels have been the

Fig. 127 Stress corrosion cracking in a bellows expansion joint of stainless steel for a district heating line; the attack has been caused by chloride-containing water dripping from the ceiling and evaporating on the hot surface of the bellows.

object of much research but are still not fully resolved. It is most likely that the rate-determining reaction step changes with the conditions. In many cases, however, local weaknesses in the passivating layer seem to play an important role. Thus the risk for stress corrosion cracking is often related to the electrode potential values which correspond to instability of the passivation on the anodic polarisation curve (Fig. 128).

The risk for stress corrosion cracking can be counteracted by various methods, e.g.:

— *stress-relief annealing,* usually at temperatures immediately below the critical temperature for sensitisation, i.e. about 500°C,
— *shot peening*, whereby compressive stresses are introduced into the outer zone of the material,
— design so that stress concentrations are avoided, and
— choice of a stress corrosion resistant alloy.

Stress corrosion testing can be carried out by exposure of loaded or cold-deformed test pieces, for example:

— in an autoclave with highly pure water; for the testing of the resistance towards intergranular stress corrosion cracking under conditions similar to those in a BWR reactor,
— in a 40% $CaCl_2$ solution at 100°C; for the testing of the resistance towards transgranular stress corrosion cracking in media containing a high chloride concentration; testing was previously carried out in a boiling 45% solution of $MgCl_2$ at 154°C, but this test has been found too severe, and
— in an evaporating, dilute NaCl solution; in which the objective is to simulate an ordinary case of transgranular stress corrosion cracking as may be caused by water containing chloride evaporating on a hot stainless steel surface.

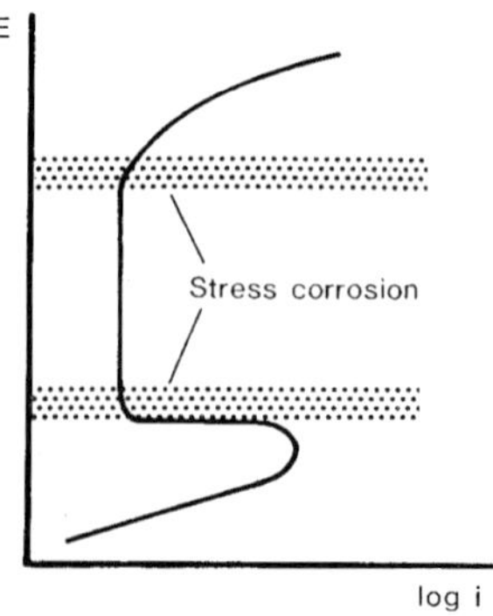

Fig. 128 Anodic polarisation curve for stainless steel; in the shaded potential regions there is an especially great risk for stress corrosion cracking.

8.3 ALUMINIUM AND ITS ALLOYS

8.3.1 Materials

Aluminium is something of a latecomer as a construction material. A small quantity of the element was produced for the first time in 1825 by the Dane H. C. Örsted, but it was not until 1886 that the Frenchman Héroult and the American Hall working independently invented a process for the technical production of aluminium. At the beginning of the twentieth century aluminium was therefore a rather rare metal. At court it was equally elegant to eat from an aluminium plate as from one made of gold. Nowadays aluminium is one of the most common metals. It is used in the unalloyed form as well as alloyed with other metals for a great number of important applications, e.g.:

— Al 99.0–99.7† for cooking and other household utensils, foil and other equipment,
— AlMn1 for roofs and façades,
— AlMn1Mg1 for beer and lemonade cans,
— AlMgSi for rods and sections; by thermal ageing a precipitation-hardening is achieved which increases the strength of the material,
— AlZn5Mg1 for constructions requiring high strength including welded structures, e.g. bridges and cranes; the strength is improved by natural ageing, and
— AlZnMgCu alloys where high strength is demanded, e.g. in the aircraft industry.

8.3.2 General corrosion characteristics

Aluminium is a *base (active) metal* having a strong tendency to react with its surroundings. This means that an aluminium surface which is exposed to the air will rapidly acquire a thin coating of aluminium oxide (about 0.01 μm) which protects it from further attack. Without this effective *passivation* aluminium could not be used in practice. The corrosion characteristics of aluminium are shown in the potential-pH diagram (Fig. 129).

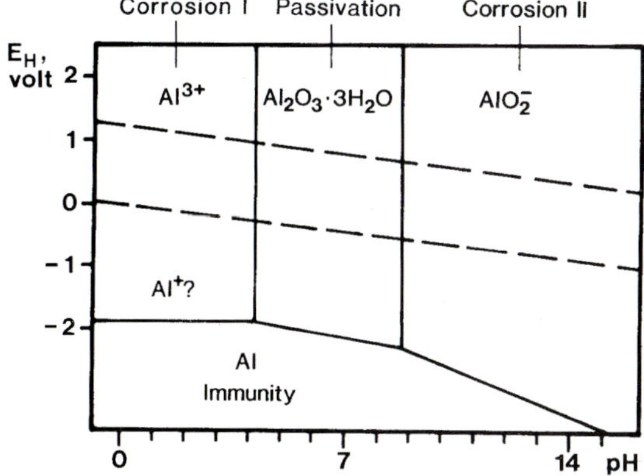

Fig. 129 Potential-pH diagram for Al–H_2O at 25°C; 10^{-6} M dissolved Al [2].

† The figures represent the percentage content of the element in the alloy.

The corrosion properties are to a large extent the same for the unalloyed metal as for most of the conventional aluminium alloys. The high-strength materials, e.g. A1Zn5Mg1 and the copper-containing alloys are exceptions since they have a greater tendency to undergo corrosion.

8.3.3 Uniform corrosion

Uniform corrosion of aluminium in outdoor atmospheres is usually negligible. The average penetration for A1Mn1 during exposure for 20 years does not exceed 1 μm/year, even in a polluted atmosphere, and in a clean atmosphere it is much less.

Solutions having a pH value outside the passivation domain of the potential-pH diagram, i.e. acidic or alkaline solutions having pH<4 and >9 respectively, can, however, cause rapid uniform corrosion of aluminium materials. Therefore, cooking vessels of unprotected aluminium should not be used for acidic food in order to avoid corrosion damage to the vessel as well as contamination of the food with aluminium; it is uncertain whether such contamination may have adverse health effects.

Further, freshly prepared mortar is alkaline and therefore corrosive towards aluminium. In order to avoid the development of etched regions on the metal surface care should be taken to avoid mortar splashes. During building work the aluminium parts can be covered with plastic sheeting or with a strippable lacquer or tape. Aluminium surfaces which are in contact with fresh concrete, e.g. barge-boards or window sills, are certainly etched at first but soon get a coating of calcium aluminate which protects against further corrosion. Corrosion damage can occur, however, if the concrete is porous or the construction design is such that the aluminium surface is repeatedly exposed to alkaline water from the concrete.

8.3.4 Pitting

In polluted *outdoor atmospheres* small *pits* occur which are hardly visible to the naked eye (Fig. 130). Above the pits small crusts of corrosion products are formed — usually aluminium oxide and aluminium hydroxide. The growth in pit depth is relatively rapid during the first few years but eventually stops (Fig. 131), and the depth seldom exceeds 200 μm even after several decades. This is also true in atmospheres polluted with sulphur dioxide. For AlMn1 the following values of the maximum pit depth have been measured in different types of atmosphere:

Atmosphere	Max pit depth after 20 years' exposure (μm)
Rural	10–55
Urban	100–190
Marine	85–260.

Such shallow pits do not generally influence the mechanical strength of the constructions. However, the bright new-looking appearance of the surface gradually disappears as the grey coating — the patina — of corrosion products develops. If the atmosphere contains much soot, this is adsorbed by the corrosion products and the

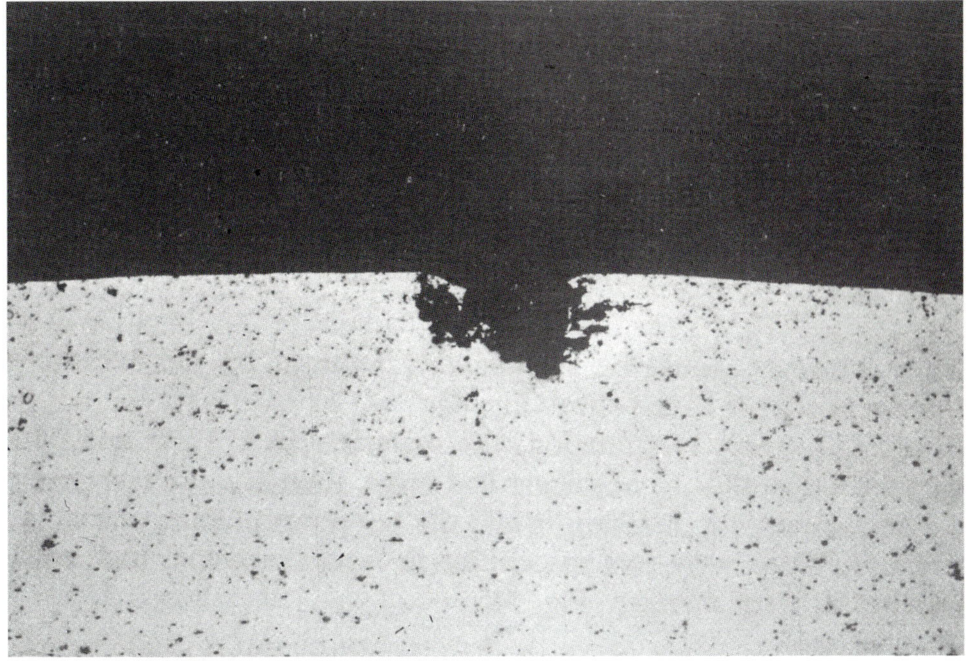

Fig. 130 Pits in aluminium after 10 years' exposure to an industrial atmosphere; bottom, surface view; top, cross-section (×60).

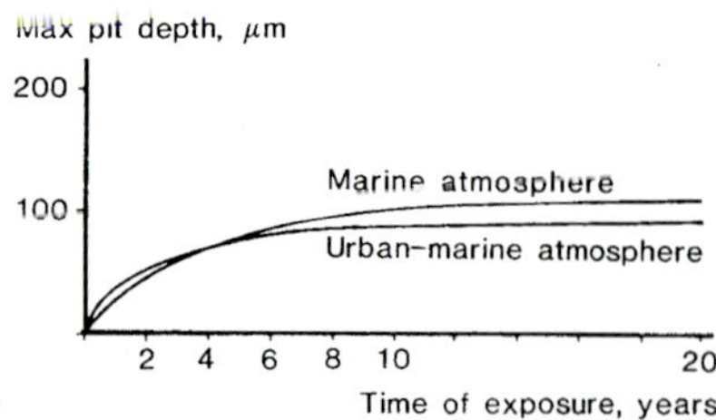

Fig. 131 Maximum pit depth in a sheet of AlMnl.2 after 20 years' atmospheric exposure in Sweden, 1955–1975 [18].

Fig. 132 Discoloration of a door of anodised aluminium protected from rain.

patina becomes dark. Aluminium on façades and other parts of buildings generally ages uniformly and with an attractive appearance if the metal is exposed to the washing action of the rain. But, conversely, the parts protected from the rain, can corrode with the formation of an uneven grey discoloration (Fig. 132). This can be countereacted by washing the surface regularly with water containing a suitable surfactant.

If the aluminium is permanently exposed to *water*, then pitting can be more serious, especially with stagnant water. The presence of oxygen and chloride, or other halide ions, determines whether, and to what extent, attack will take place. If HCO_3^- and Cu^{2+} ions are also present, then the risk of pitting will be increased. A prerequisite is that the pitting potential is exceeded. The value of this potential varies according to the composition of the material and the environment; in sea water it is of the order of –0.4 V (SHE) for most alloys.

It is chiefly the cathode reaction outside the pit, i.e. the reduction of oxygen on surface inclusions of noble material (Cu or Al_3Fe), which controls the rate of pitting.

This in turn means that the growth of the pit depth slows down as the diameter increases. It has been found in fresh water as well as sea water, that the pit depth (P) is proportional to the cube root of the time (t):

$$P = \text{const}\,\sqrt[3]{t}\,.$$

In consequence the time for perforation increases considerably with increasing thickness of the component. Since the cathode reaction controls the pitting, it is often advantageous to divide the attack (i.e. the available cathodic current) among a large number of pits; the depth of each pit will then be less than in the case of a smaller number of pits. In order to increase the density of pits, a small amount of copper (up to 0.15%) is sometimes added to the aluminium alloy.

Aluminium cannot generally be used for water pipes because of the risk of pitting in stagnant water. Nevertheless, this can be effectively counteracted by keeping the water moving, in tap-water pipes a flow rate of 2.4 m/min is sufficient. Aluminium and its alloys have been found suitable for special pipes, e.g. mobile irrigation equipment, where low weight is of importance and the exposure time is short.

Pitting can also occur at the bottom of saucepans made of aluminium. The attack can be initiated by copper deposits from tap water or from charred food residues. A special type of pitting, so-called pinholing, can lead to perforation of thin-walled household food containers within 2–3 days when these are allowed to stand with contents of acidic, jelly-like foods or salty gravy.

8.3.5 Bimetallic corrosion

Since aluminium is a base metal there is risk of *bimetallic corrosion* on direct contact with a more noble material such as copper, carbon steel or stainless steel. A prerequisite for attack is, however, that an electrolyte is present at the point of contact. Bimetallic corrosion does not therefore take place in dry indoor conditions. The risk is also low on inland outdoor exposure (rural or urban atmosphere). But, on the other hand, it has to be taken into account in a pronounced marine atmosphere (on ships or near the sea) (see Appendix 1) or when the points of contact are immersed. Under these conditions the metals should be electrically insulated from each other, e.g. with plastics (see Fig. 50), or by coating the contacting surfaces with corrosion-preventing paint.

A risk of bimetallic corrosion of aluminium can occur in outdoor atmospheres, e.g. if the metal is exposed to copper-containing water so that small copper particles are deposited and form micro-cathodes on the aluminium surface. Bimetallic corrosion can then lead to the surface becoming rough and discoloured by the corrosion products. This type of corrosion can also occur if the surface is contaminated with soot.

8.3.6 Crevice corrosion

A kind of *crevice corrosion* in aluminium can occur if water, e.g. rainwater, is allowed to collect between the sheets in a stack or between the turns of a strip coil. This crevice corrosion can result in the formation of aluminium oxide which, in the form of *water stains*, will discolour the surface. The water stains are difficult, or may be impossible, to remove.

Crevice corrosion of practical importance can also occur in crevices between aluminium surfaces where the crevice is filled with a chloride-containing medium, e.g. road salt, sea water or salt deposits from a marine atmosphere. The corrosion products can sometimes be so voluminous that they force apart the components of a structure (an example of the phenomenon of oxide-jacking).

8.3.7 Layer corrosion

Layer corrosion sometimes referred to as exfoliation corrosion, is mostly confined to rolled or extruded material of the AlCuMg and AlZnMg types. The attack is localised in thin parallel layers oriented in the direction of processing and leads to unattacked metal layers being released like pages in a book, or to blisters being formed on the metal surface (see 4.7). Layer corrosion appears on exposure to stagnant water or a marine atmosphere. The resistance to layer corrosion is mainly determined by the ageing treatment of the metal. The best resistance is usually obtained by natural ageing for AlCuMg alloys and by artificial ageing for AlZnMg alloys.

8.3.8 Stress corrosion cracking

This type of corrosion occurs on only a few types of aluminium alloys such as AlCuMg, AlMg (Mg >4.5%) and AlZnMg (with high Zn and Mg contents). For *stress corrosion cracking*, tensile stresses are required above a certain critical magnitude, but no specific corrosive agent is required. Stress corrosion cracking seems to appear even in the presence of moisture alone (Fig. 133). The rate of

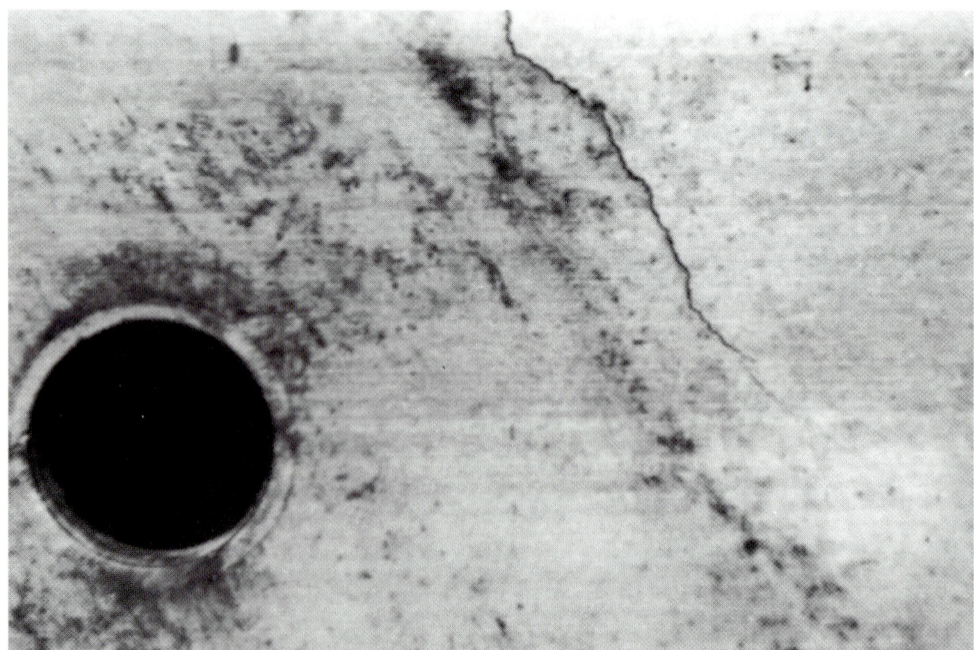

Fig. 133 Stress corrosion cracking in a tubular support of AlZn5Mg1 due to unsuitable rapid cooling after heat treatment (Gränges Aluminium).

cooling after thermal treatment influences the resistance to stress corrosion cracking; in AlZnMg slow cooling usually favours resistance to this form of attack although the reverse is true for AlCuMg alloys.

8.3.9 High temperature oxidation

In dry air an aluminium surface is rapidly covered with a thin oxide film which protects against further oxidation. The thickness of the layer even after a long time at 500°C reaches only 0.1–0.2 μm. Aluminium is therefore often used in the construction of furnaces.

8.3.10 Cathodic protection

Aluminium materials in water can be protected against pitting by *cathodic protection*. Here the electrode potential is reduced to a value below the pitting potential of the material in the environment concerned. Hydrogen gas can, however, be produced at the cathode and this will result in a rise in the pH value. If the pH rises too much then the aluminium can be attacked. *Overprotection* should therefore be avoided, i.e. the potential should not be reduced below a certain critical value; in soil and fresh water this value is –1.2 V (against the copper/copper sulphate electrode). In practice aluminium can in many cases be protected with the aid of sacrificial anodes; zinc or AlZn5 anodes for constructions in sea water; magnesium anodes for constructions in fresh or brackish water as well as for unpainted surfaces underground, and zinc anodes for painted structures underground. Cathodic protection can also be achieved by cladding with a metal which is less noble than the substrate, e.g. a coating of AlZn1 on a substrate of non-alloyed aluminium.

8.3.11 Anodising

The oxide layer formed on an aluminium surface on exposure to air provides good corrosion protection, but this oxide layer can be made thicker by electrolysis. The treatment is called *anodising* and the oxide layer so formed an *anodic oxide coating*. Resistance to corrosion is increased by anodising with the 'new' appearance of the surface being maintained on exposure to outdoor atmospheres. The anodic oxide coating also protects against wear and is an electrical insulator. Anodised aluminium is chiefly used in the building industry, e.g. for façade cladding and window frames, but also in other areas, e.g. for masts, booms and fittings on sailing boats (Fig. 134).

When anodising aluminium the metal object is made the anode in an electrolytic cell. The electrolyte usually consists of a solution of sulphuric acid, sometimes with the addition of organic acids. The anodic oxide coating formed during electrolysis consists of a compact barrier layer near the metal surface, and above this a layer containing micropores (Fig. 135).

The anodised layer can be *coloured* by immersion of the object in a dyeing solution. The colouring agent is then enclosed in the micropores. Even the dyeing process can be carried out with the aid of electric current. Agents having good colour fastness should be chosen for colouring otherwise undesirable colour changes can occur on the anodised surface.

A coloured anodised surface can also be obtained directly during the anodising

Fig. 134 Masts, booms and fittings on sailing boats are often made of anodised aluminium.

Fig. 136 Anodised aluminium in different colours.

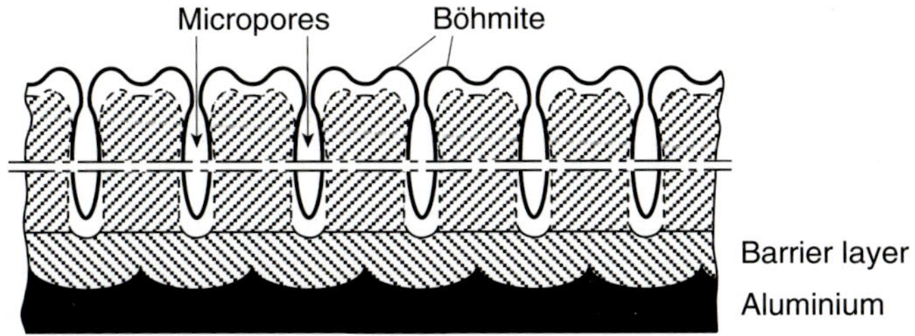

Fig. 135 Cross-section showing the structure of an anodic oxide coating on aluminium; before (- - - -) and after (———) sealing.

process. Thus, grey coloration is formed on aluminium alloys having a high silicon content. This is because particles from the substrate are enclosed in the anodic oxide coating. Furthermore, different bronze and black colours can be obtained directly by anodising in electrolytes containing various additives, e.g. sulphosalicylic acid (Fig. 136).

Irrespective of which anodising method is used or if coloration is carried out, anodising must be completed by *sealing*. This is usually done by immersion in boiling de-ionised water and results in crystallisation of the outer part of the coating. Crystallisation leads to an increase in volume which in turn causes the micropores to close up. Sealing provides a considerable increase in the corrosion resistance of the anodic oxide coating.

The protective power of the coating is generally proportional to its thickness. For outdoor use the recommended thickness of the coating is at least 15 and preferably 20 μm, and for severe conditions 25 μm. For indoor use 5–10 μm is usually sufficient.

8.3.12 Painting

Aluminium material for outdoor use, on buildings for example, does not generally need corrosion-preventing painting for technical reasons; as previously mentioned, atmospheric corrosion is not usually so extensive that the strength of the construction is significantly affected. However, painting of aluminium is carried out to a large extent for aesthetic reasons in order to give colour to the surface or to maintain its original new appearance.

When painting aluminium it is important that the surface is prepared in a proper way. After de-greasing, a process of chromating, phosphating, anodising or the application of etch primer is required to provide good paint adhesion. Further coats are subsequently applied consisting of primer and top coat chosen with respect to the required corrosion resistance, colour durability, plasticity etc.

There are coil-coated aluminium sheet products on the market having different types of top paint, e .g. alkydmelamine, polyester or polyvinylidenefluoride of which, the last usually has the best technical characteristics, but is also the most expensive.

Painting of aluminium can also be carried out on the building site by brushpainting, where the operation consists of the following steps:

— cleaning and degreasing, e.g. with white spirit,
— the application of etch primer,
— the application of primer; necessary only under extremely corrosive conditions, and
— the application of air-drying top paint of, for example, a vinyl or urethane type designed for out-of-door use.

Another alternative is to apply a solution of bitumen after degreasing the surface with white spirit. The thickness of the bitumen coating should be at least 50 μm.

8.4 COPPER AND ITS ALLOYS

8.4.1 Materials

Copper can be found in the metallic state in certain geological formations, so-called native copper. The metal copper has been used by human beings for more than 10 000 years. It is used unalloyed, as well as alloyed with other metals such as zinc (brass), zinc+aluminium, tin or nickel (special brass), tin (phosphor bronze), tin+zinc+lead (gun metal), aluminium (aluminium bronze) and nickel (copper-nickel).

Unalloyed copper is often used for roofing, especially for monumental buildings, and sometimes for façade cladding.

Copper is also used extensively in the conveyance of water. For example, phosphorised copper is used for hot and cold water piping in residential buildings and in water heaters. Various types of brass are used for fittings in tap water supply and in heating systems. Aluminium brass and copper nickel are common tubing materials in condensers and other heat exchangers, e.g. in heat pumps and in desalination plants for sea water. Aluminium bronze is used in valves and pumps for sea water.

Other important uses are in electrical equipment including electronic devices, because of copper's good electrical conductivity.

8.4.2 General corrosion characteristics

Copper is a comparatively *noble metal*. Its immunity domain in the potential-pH diagram reaches some way into the stability domain of water (see Fig. 15). But in oxygen-containing water, corresponding to the upper parts of this area, there is risk of corrosion. The layer of corrosion products which sometimes forms on the surface does not always provide effective passivation, although basic copper salts that are present may provide a degree of protection.

Corrosion damage of copper materials exposed to aqueous corrosion media is usually of the uniform type although pitting is not unusual and can present problems. The average penetration resulting from uniform corrosion is often small. The following guiding values can be given for copper in various environments; these also apply to several copper alloys with the exception of zinc-rich brass (having more than 20% zinc), which can show higher values as a consequence of dezincification:

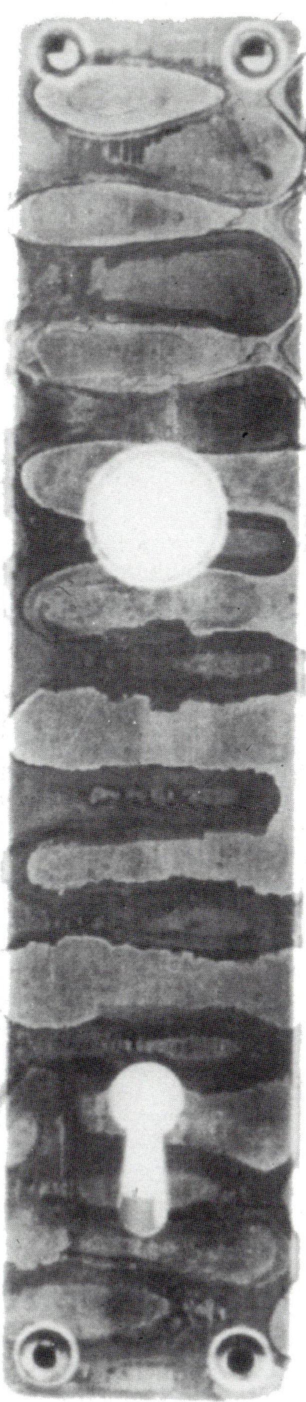

Fig. 137 Water stains on brass.

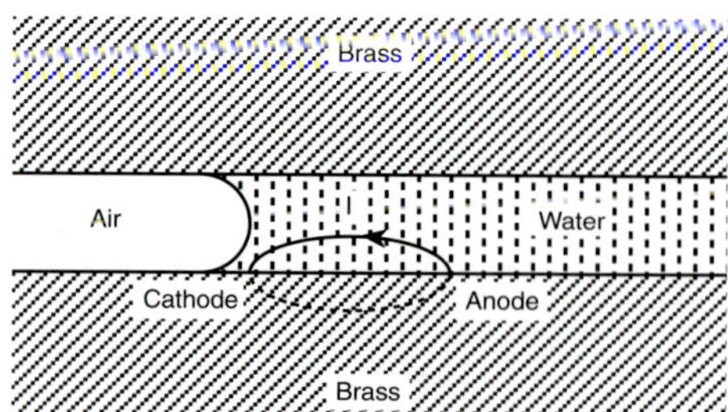

Fig. 138 Aeration cell in a partly water-filled crevice between two brass plates; the cell action causes discoloration of the brass.

Atmosphere	0.5–2.5 μm/year
Fresh water	about 10 μm/year
Sea water	about 50 μm/year
Soil	1–50 μm/year.

8.4.3 Discoloration

Copper materials are sensitive to discoloration. Just a little water is required in the crevices between stacked sheets or between the turns of a coil for *water stains* to appear after only a few days (Fig. 137). This is because of the action of differential aeration cells in the enclosed water (Fig. 138). The metal surface nearest the contact area of the water with the air acts as a cathode and here oxygen is reduced to hydroxide ions. The water soon becomes depleted in oxygen at longer distances from the air. Anodic oxidation takes place here with the formation of dark copper oxide which discolours the surface. Water stains on surfaces facing each other show corresponding but reversed patterns. In order to avoid water stains copper material should be protected from rainfall and condensation during storage and transportation. Condensation can occur when cold material is taken into a warm environment since the temperature then easily falls below the dewpoint.

Another type of discoloration is called *salt stains* because it arises from the presence of hygroscopic salts on the surface. These can result from the evaporation of residues of washing water used after pickling processes. In the solution drops formed around the hygroscopic salt crystals, differential aeration cells develop, and a dark, discolouring copper oxide results.

Fingerprints also carry hygroscopic salts; after a few days fingerprints appear as a dark discoloration. When handling copper one should therefore use gloves if surface discoloration is to be avoided.

Furthermore, sulphide, e.g. in the vapours from the boiling of fish, can cause discoloration of copper objects in the vicinity.

8.4.4 Patina formation

A newly laid copper roof usually has a patchy appearance at the start, often to the owner's disappointment. But after one or two years the surface generally receives an even, dark colour. This stage lasts for a few years. Later a green coloration, a so-called *green patina* appears (Fig. 139). In an urban or industrial atmosphere

Fig. 139 Two copper-covered towers in Stockholm; the tower to the left with a dark oxide coating after a few years' exposure; the tower to the right with green patina after several decades of exposure.

containing a moderate content of sulphur dioxide, and in a marine atmosphere, the green patina usually begins to appear after about seven years. But in a clean rural atmosphere it can take tens or even hundreds of years. This is because the green patina generally receives its colour from copper hydroxide sulphate, which under marine conditions is in a mixture with copper hydroxide chloride. These salts are in fact corrosion products. In order for the green patina to form, a supply of SO_2 and Cl^- respectively is necessary, as well as a pH value in the moisture film on the metal surface which is not too low (see Fig. 16). The green patina is formed more slowly on vertical than on horizontal or sloping surfaces, since the former are wet for a shorter

Fig. 140 Downpipe of cast-iron, which has been attacked by bimetallic corrosion; the pipe has led water from a copper roof and dissolved copper has been deposited on the iron and caused heavy rusting.

period of time. When the moisture on the metal surface has a low pH value, e.g. in the vicinity of chimneys, copper hydroxide sulphate cannot be precipitated. Rain draining from copper surfaces in these situations has, as a consequence of its copper content, a tendency to produce a blue-green discoloration on masonry, stone etc. It should therefore be led away by gutters and drainpipes, preferably of copper or plastics. Steel pipes may be attacked by bimetallic corrosion, due to deposition of copper from the water (Fig. 140). Aluminium should never be used for this purpose. Green patina on roofs and façades of copper can also be achieved artificially. Many methods for artificial patination of copper have been suggested, but the failures have been numerous. One method, which has been successfully used, is based on the application of copper hydroxide salt on the copper surface with appropriate treatment of the surface before and after the application.

8.4.5 Copper-containing water

The use of copper piping for hot and cold tap water has generally been successful. But in exceptional cases inconvenience has been caused by corrosion. One corrosion effect is the appearance of copper traces in the water. This phenomenon is called cuprosolvency, and, in extreme cases, the copper-containing water is called "blue water".

In a new copper pipe installation the water may contain about 1 mg Cu/litre, even after a few minutes' running. After standing overnight the copper content can be even higher. But with time a protective layer of copper hydroxide carbonate and calcium carbonate is usually formed on the pipe walls, resulting in the level of copper in the water falling to a few tenths of a mg Cu/litre after some months. Under unfavourable conditions, for example low pH value in the water, a higher content of dissolved copper can occur. A high copper content can give the water an unpleasant taste and give rise to a blue-green discoloration in washbasins and baths. Damage can also be caused to the laundry by the discoloration and breakdown of textile fibres. Furthermore, the traces of copper may cause bimetallic corrosion in aluminium vessels and pipes of galvanised steel exposed to the water. Another adverse effect is copper contamination of waste water and of the sludge obtained in waste water purification.

There has been much discussion around the question of how human beings are affected by copper contained in drinking-water from copper piping. According to the World Health Organisation (WHO) such low copper concentrations are not generally a health hazard. On the contrary, some intake of copper is necessary for the metabolism. Still a relation has been suspected between high copper content in the drinking-water and diarrhoea with infants. With regard to the 'aesthetic quality' of water the WHO has recommended a guideline value of max. 1.0 mg l^{-1} Cu.

Water showing a tendency towards unusually high dissolution of copper can be made less cuprosolvent by raising the pH value; a pH of about 8 should be aimed at. It is also recommended to flush for one or two minutes after standing overnight, before taking water for drinking or cooking from a copper pipe.

Copper which is dissolved can also have certain advantages. The low copper concentration in tap water has been shown to exhibit a bactericidal action. Further, copper released from copper material in sea water hinders the growth of fouling organisms. This is of great advantage, for example in brass or copper-nickel heat exchangers and on copper-nickel sheathed parts of offshore platforms.

8.4.6 Pitting

Uniform corrosion in tap-water pipes made of copper does not damage the pipe in the sense that it becomes unfit for service. More detrimental to the pipes is *pitting* (Fig. 141) which can occur in unfavourable conditions. Three types of pitting have been described:

— *Type I* can occur in cold-water pipes of annealed or half-hard copper carrying water from a ground-water supply. It is caused by an interior carbon film in the pipe, formed by cracking of residues of drawing lubricant used in the final

Fig. 141 Pit (with and without corrosion products) in a copper pipe, caused by corrosive water at about 65°C (Metallverken).

annealing of the pipe. Annealed or half-hard copper piping should not be allowed to have more than 0.2 mg carbon per dm^2 of the inner surface and the carbon film which is present should not be continuous.

— *Type II* appears in hot tap-water piping. It is caused by unsuitable water composition; a pH value below 7.4 and a low HCO_3^-/SO_4^{2-} ratio (less than 1).

— *Type III* is rare but has occurred in cold-water lines made of hard as well as annealed copper pipes. The main reason is probably that the water has such a low HCO_3^- concentration that no protective coating of copper hydroxide carbonate is formed on corrosion. Increase of the HCO_3^- concentration to at least 70 mg l^{-1} seems to be an effective countermeasure.

Pitting of copper used in water supplies may occur due to other causes that are not yet fully understood. There is evidence that pitting may occur under silt deposits and even under biofilms as a result of microbial activity (MIC).

8.4.7 Erosion corrosion

Copper materials are relatively sensitive to *erosion corrosion* when they are exposed to water with high flow velocity, especially where the flow is disturbed so that turbulence occurs. The attack has a characteristic appearance; broad pits, free from corrosion products, sometimes horseshoe-shaped, and (seen in cross-section) undercut in the direction of flow (Figs 142 and 143) (cf. Figs 34 and 35).

Fig. 142 Attack by erosion corrosion in a copper water pipe (Metallverken).

Fig. 143 Horseshoe-shaped attack by erosion corrosion in copper water pipe; direction of flow from left to right.

Table 12 Highest recommended flow velocities of fresh water in copper distribution lines in buildings, according to Swedish norms

Type of pipe line	Highest recommended flow velocity (ms^{-1})	
	Cold water	Hot water
Non-exchangeable circulation lines (continuous flow)	0.8	0.6
Exchangeable circulation lines (continuous flow)	2	1.5
Non-exchangeable distribution lines (intermittent flow)	2	1.5

Table 13 Highest recommended flow velocity of sea water in pipes of different materials; the values are valid for permanent flow

Pipe material	Flow velocity (m s^{-1})
Phosphorized copper (Cu-DHP)	1.0
Admiralty brass (CuZn28Sn1As)	1.8
Aluminium brass (CuZn20Al2As)	2.2
Copper nickel (CuNi10Fe1Mn)	2.5
Copper nickel (CuNi30Mn1Fe)	3.5

To avoid the risk of erosion corrosion copper piping must be dimensioned so that the flow velocity does not exceed a given limit, bearing in mind consideration of the permanency of flow, water temperature and accessibility of the pipe (Table 12). It is also important that the pipe is streamlined without unnecessary turbulence at joints and bends. Guidelines on the highest recommended flow velocities, for heat exchanger pipes made of different copper alloys are available (Table 13).

8.4.8 Dezincification

The dezincification of brass is the selective dissolution of the zinc from the alloy leaving a porous mass of copper having poor strength (Fig. 144) (cf. Fig. 30).

Fig. 144 Cross-section through a stem of (α+β)-brass, used in a water valve; the stem has been dezincified; in the threads there are white corrosion products which often cause the stem to stick.

Brass can be dezincified by contact with water as well as exposure to an outdoor atmosphere. The attack is accelerated by the presence of chloride and by raised temperatures. Two main mechanisms for dezincification have been presented:

— total dissolution of brass at the surface followed by redeposition of dissolved copper, and
— selective dissolution of zinc from the surface leaving copper behind; in this case dezincification can proceed due to volume diffusion of zinc from the bulk of the brass to the surface.

The resistance to dezincification in brass increases with the copper content (brass with a copper content of above 63% consists of a single phase — α-phase) and brass having more than 85% copper is practically resistant to dezincification even in sea water. Even α-brass having a lower copper content can be made resistant to dezincification by the alloying addition of an *inhibitor*, usually 0.02–0.04% arsenic, antimony or phosphorus. Condenser pipes are often manufactured from aluminium brass or admiralty brass, both of which are α-type brass with an arsenic additive.

The inhibitor is not, however, effective in the β-phase of $(\alpha+\beta)$-brass, i.e. brass (having a relatively low copper content) which is composed of α- and β-crystals. Such brass is used in water-pipe fittings, because of its good characteristics in the various manufacturing operations, such as extrusion, die casting, hot forging and cutting processes. Swedish norms require that brass used in fittings in residential buildings must be resistant to dezincification since this type of corrosion can lead to valve stems sticking or breaking, to blocking of the pipes with corrosion products, or to leakage. The resistance to dezincification in $(\alpha+\beta)$-brass requires not only the presence of arsenic but also that the β-phase occurs in isolated grains in the α-phase, or that the brass has such a composition that, at raised temperatures (as arise in extruding or hot forging), it is an $(\alpha+\beta)$-brass, but, at the working temperature of the fitting, an α-brass. One can also add tin and aluminium to the alloy to increase its corrosion resistance.

In the ISO 6509 standard a test method is described for determination of the dezincification resistance of brass. In this test the depth of dezincification is measured after the test piece has been exposed to a 1% copper chloride solution at 75°C for 24 h.

8.4.9 Dealuminisation

Aluminium bronze having more than about 8% Al has very good strength characteristics and in addition good corrosion resistance, provided that the alloy is free from the aluminium-rich γ_2-phase. This phase is very sensitive to selective corrosion, i.e. to *dealuminisation.* In order to decrease the risk of γ_2-phase arising in the material one should ensure suitable conditions on heat-treatment and welding. The risk can also be counteracted by adding nickel, iron and manganese to the alloy. Nickel-aluminium bronze is a strong and corrosion resistant material, which has been found suitable for marine applications, e.g. propellers, valves and tube sheets in heat exchangers.

Fig. 145 Stress corrosion cracking in a brass kerosene stove; the attack has been caused by residual stress from the manufacture of the brass container in conjunction with ammonia-containing soldering flux remaining from the attachment of the legs.

8.4.10 Stress corrosion cracking

Stress corrosion cracking in copper materials is caused by tensile stresses, usually residual stresses from cold working, in combination with a corrosion environment which contains ammonia+moisture, mercury or related substances. Examples of such media are ammonia-containing solder fluxes, urine, atmospheres in stables and even outdoor atmospheres (Fig. 145). Since the risk of cracking is greatest during the seasons when the humidity is high, the phenomenon is sometimes called *season cracking*. Other substances, e.g. nitrite, have also been shown to cause stress corrosion in copper materials. The cracks can be transgranular or intergranular depending on the pH value of the environment and the magnitude of the stresses.

It is generally the zinc-rich brasses which are most susceptible to stress corrosion but under special conditions other copper materials, even pure copper, can be damaged by this type of corrosion.

Often an effective countermeasure is to remove the residual mechanical stresses from the construction by *stress-relief annealing*; typically for brass at 275-325°C for 1–2 hours (Fig. 146). Inhibitor treatment, corrosion-preventing painting, tin, nickel, or chromium plating do not generally provide satisfactory protection against stress corrosion cracking.

The resistance of a copper alloy object to stress corrosion cracking can be determined according to ISO 6957, whereby the object is exposed to an ammonia-containing atmosphere for 24 h.

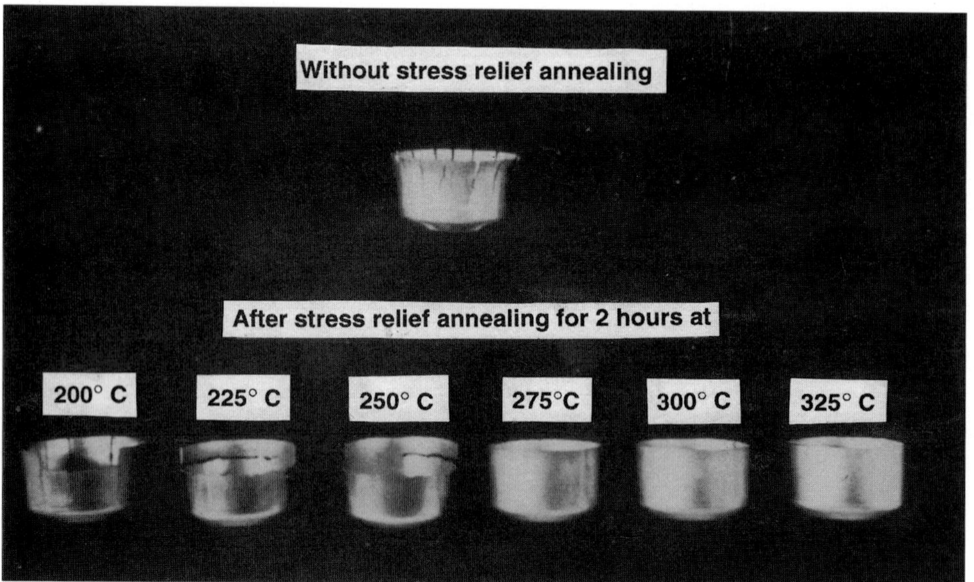

Fig. 146 The influence of stress-relief annealing on the stress corrosion resistance of deep drawn brass cups (Cu63Zn37); after the annealing the cups were exposed to an ammoniacal testing solution.

8.4.11 Corrosion fatigue

Cracking as a result of *corrosion fatigue* can occur, for example, in hot-water piping of copper if it is mounted too rigidly, i.e. without sufficient flexibility. The variations in length due to changes in temperature can cause fatigue which is reinforced by corrosion, and can lead to cracking of the pipe. The cracks are as a rule wide and branched (see Fig. 45) and are predominantly transgranular. An effective countermeasure is to provide the pipe with an expansion joint to accommodate the expansion, e.g. a horseshoe, loop, gland or bellows (see 4.11.2).

8.5 TITANIUM AND ITS ALLOYS

Titanium in the metallic state was first produced by Berzelius in 1825. Not until the middle of the 20th century, however, the metal could be produced in such quantities and at such a price that any significant technical application was possible. Several types of titanium materials are now available, e.g. unalloyed, palladium alloyed, aluminium-vanadium alloyed and molybdenum-nickel alloyed titanium (ASTM B 265).

Like aluminium, titanium is a base (active) metal, with a strong tendency to react with other species in its environment. Exposed to air a titanium surface will rapidly acquire a thin coating of titanium oxide, which protects from further attack, and heals, if scratched. The passivating layer formed is stable within a wide pH range, but can fail in acidic and reducing conditions. Due to their passivating properties, the titanium materials are very corrosion resistant in most environments. It is remarkable that the oxide coating is protective even in moist chlorine and in chloride-containing media, which often cause localised corrosion on other passivated metals. Therefore, the titanium materials are used, e.g. in plants for chlorine bleaching within the pulp industry, in cooling systems using sea water, and in desalination plants for sea water. In spite of comparatively high investment costs, equipment made of titanium is often profitable owing to its good corrosion resistance.

Titanium, however, cannot be a solution to all corrosion problems. The metal may be attacked by, e.g., dry chlorine gas, hydrofluoric acid, acidic fluoride solutions, hydrogen peroxide solutions with more than 5% H_2O_2, anhydrous methanol and reducing acids free from corrosion inhibitor. It should also be mentioned that titanium in the passive state behaves like a fairly noble metal, which may cause bimetallic corrosion on less noble metals.

9
The methodology of corrosion investigations

Corrosion investigations are carried out in many contexts, e.g. for:

— the development of new materials and corrosion-preventing agents,
— the selection of construction materials,
— the quality control of materials and protective agents,
— corrosion monitoring, and
— analysis of corrosion failures.

In addition to the conventional methodology for chemical analysis, metallographic investigations and tensile testing special methods are used for exposure testing and corrosion monitoring as well as for electrochemical and physical surface investigations. These methods will be briefly described here.

9.1 CORROSION TESTING

Two main types of corrosion testing can be distinguished: in-service testing and laboratory testing.

9.1.1 Corrosion testing under service conditions

The most representative and reliable results are obtained if the test piece is exposed to conditions which are as near as possible to the service conditions. This can be done by field tests or by service tests.

In *field tests* the test piece is exposed to the same type of atmosphere, water or soil as that occurring where the material is to be used (see 5). The results from well conducted field tests are generally judged to be reliable. One should, however, note that test panels on a test site are exposed to somewhat different conditions from those, for example, in the proximity of a façade cladding on a building, where drying, rain protection, slope etc. will not necessarily be the same.

In *service tests* the test piece is exposed in a piece of industrial process equipment, for example, under the actual service conditions prevailing.

9.1.2 Corrosion testing in the laboratory

In field tests as well as service tests the service conditions vary, often in a manner difficult to define and control. It can therefore be advantageous to carry out exposure testing in the laboratory *under simulated service conditions*, which can be better controlled. But there is always the risk that some important factor is overlooked which in service may be critical for corrosion. Thus, with corrosion testing in sea water a lower rate of corrosion has often been found in artificial sea water, prepared e.g. according to ASTM D 1141–75, than in natural sea water. This difference is probably due to natural sea water containing low concentrations of compounds and microorganisms not present in the artificial sea water.

Field tests and service tests often require long exposure times. This is unacceptable in many cases, so some form of *accelerated corrosion testing* must be used. Accelerating the tests should take place via moderate changes in, for example, temperature, time of wetness or the content of corrosive agent. The testing conditions must not differ so much from those in service that the corrosion mechanism and corrosion products will differ in the two cases. Then the accelerated testing may give misleading results.

There are numerous laboratory methods used for corrosion testing, of which many have been standardised. Several of them are referred to in this book. In Table 14 some methods for accelerated corrosion testing have been compiled. The different methods of *salt spray testing* are extensively used but in many cases give results which are misleading for the use of the material in practice. *Scab-testing* has in general been found more representative. In this case the test piece is exposed out of doors and sprayed twice a week with a 5% sodium chloride solution and allowed to dry between the applications. Material which is to be used in an urban or SO_2- polluted industrial atmosphere can suitably be tested in a *climatic chamber* with an atmosphere containing low concentrations of SO_2, <1 ppm (Fig. 147); *Kesternich testing* is carried out at high SO_2 concentrations but often gives misleading results.

9.1.3 Evaluation

Evaluation of the corrosion attack after exposure of the test piece can take place by, e.g. inspection, measurement of corrosion (pit) depth or changes in strength (UTS). Usually, however, evaluation is carried out by determination of mass loss after the corrosion products have been removed by acid cleaning, where the cleaning solution used will depend on the nature of the metal. The cleaning operation is repreated a number of times and the mass loss determined in between. The results are recorded in a diagram (Fig. 148), whereby two lines AB and BC are obtained; BC represents the mass loss due to dissolution of the metal after the corrosion products have been removed. The mass of the corrosion products corresponds approximately to point D obtained by extrapolation of line BC to the intersection with the Y-axis. The mass loss of the test piece during exposure is generally recalculated as an average corrosion depth.

In corrosion testing, three or four test pieces of the same material are exposed simultaneously under the same conditions, i.e. in *replicate tests*. When the corrosion effects, for example, the mass losses or average corrosion depths of the replicate test pieces, have been determined then calculation of the *mean value* and *standard deviation* are carried out.

Table 14 Some methods for accelerated corrosion testing

Type of test	Exposure conditions
Humidity test, e.g. SEN 431603 and 431604 and DIN 50016 and 50017	Exposure to an atmosphere with high relative humidity; possibly alternating wetting and drying; possibly UV-radiation
Neutral salt spray test (NSS), SS-ISO 3768	Continuous exposure to a spray of 5% sodium chloride solution; pH6.5–7.2; 35°C
Acetic acid salt spray test (ASS), SS-ISO 3769	Continuous exposure to a spray of 5% sodium chloride solution with an addition of acetic acid; pH3.1–3.3; 35°C
Copper-accelerated acetic acid salt spray test (CASS), SS-ISO 3770	Continuous exposure to a spray of 5% sodium chloride solution with an addition of acetic acid and cupric chloride; pH3.1–3.3; 50°C
Scab test, SS 117211	Specimens exposed outdoors and sprayed with a 5% sodium chloride solution twice a week; they are allowed to dry in between successive spraying
Climatic chamber test in an SO_2-containing atmosphere	Exposure to an atmosphere with a low SO_2 content (<1 p.p.m.), with or without condensation
Kesternich test, ISO 6988	Exposure to an atmosphere with a high SO_2 content (up to 0.67%) with periodic condensation
Corrodkote test, SS-ISO 4541	Application of artificial road-mud, i.e. a suspension of clay in an aqueous solution containing ammonium chloride, ferric chloride and copper nitrate; after application of the road mud the specimen is exposed in a humidity cabinet; relative humidity 80–90%; 38°C

By *significance analysis* one can find out whether the difference between the average values for the different variants is statistically significant, i.e. is equivalent to a real difference, or is caused by occasional variations. There are different methods for such an analysis, e.g. the so-called t-test.

Furthermore, one can determine by *regression analysis* the equation for the relationship between two variables from the measured values of the variables. A linear relationship, a so-called regression line, is looked for which will indicate a close connection between the two variables, e.g. mass loss and time. By multivariate analysis one can evaluate the influence on corrosion of a great number of variables in the environment.

The equations for the statistical calculations mentioned can be found in handbooks on statistics. The calculations can be advantageously carried out with the aid of

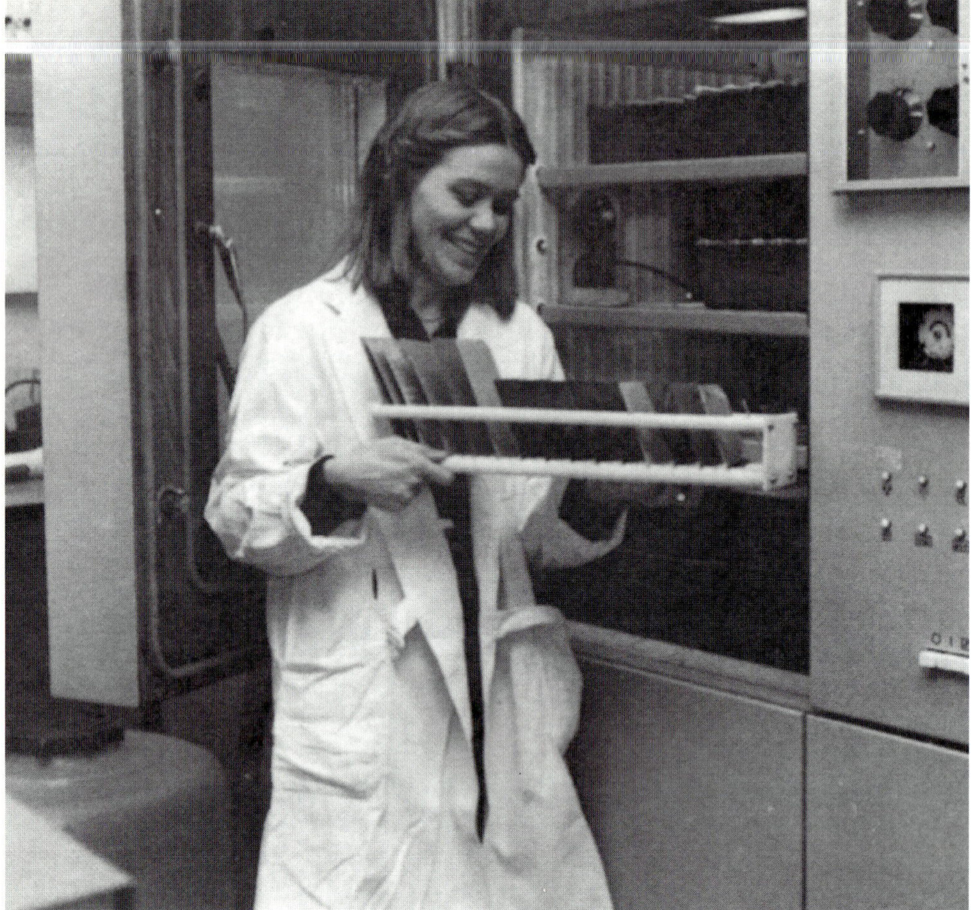

Fig. 147 Exposure of test pieces in a climatic chamber, where it is possible to add a low concentration of SO_2 to the atmosphere.

a programmable pocket calculator or with a computer; simple standard programs are available.

It is important that at the planning stage a sufficiently large number of replicate tests is included, so that statistically reliable results can be obtained.

9.2 CORROSION MONITORING

Corrosion monitoring implies that one follows the corrosion of, or the corrosivity in, a system, for example a cooling or heating system during the period of operation. As a rule the aim is to obtain information on when corrosion protection measures need to be taken.

There are several methods used in corrosion monitoring, e.g.

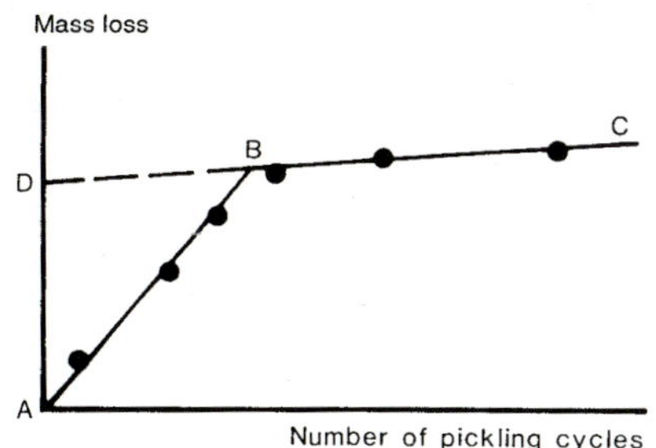

Fig. 148 Mass loss as a function of the number of cleaning cycles on pickling of exposed test pieces; the mass of corrosion products is approximately equivalent to AD (according to ISO 8407).

— *Measurement of redox potential.* The corrosivity of a liquid is usually related to its redox potential. In order to measure this one can use a probe containing a reference electrode and an electrode of an inert metal, e.g. platinum.
— *Measurement of electrode potential.* The electrode potential of a metal structure can provide information on whether corrosion is going on. For the measurement of the electrode potential a reference electrode is required.
— *Determination of mass loss.* Test coupons are exposed to the actual system and the mass loss determined after a certain exposure time. This method is simple and does not require expensive equipment (Fig. 149). But on the other hand it takes a relatively long time to determine changes in corrosivity.
— *Resistance measurements.* Test pieces, usually in the form of wire, are exposed to the actual system and their electrical resistance followed with time (Fig. 150). When the cross-sectional area of the wire is reduced as a result of corrosion, the electrical resistance increases. This method gives quicker response than the mass loss method; an indication of a change in corrosivity can be obtained within 24 h. Instruments designed for this purpose are produced commercially.
— *Polarisation resistance determination* (see 9.3). In this case an electrochemical measuring probe having electrodes made from the metal under study is inserted into the system (Fig. 151). This method gives a rapid response to changes in corrosivity; often within an hour.
— *Measurement of an acoustic emission.* This method is based on the fact that certain corrosion processes, e.g. stress corrosion cracking, emit a characteristic sound, which can be recorded and provide information about the ongoing corrosion.
— *Ultrasonic examinations.* With this technique changes in the wall thickness of pipes and tanks, for example, caused by corrosion of the inner surfaces can be followed from outside the equipment.

9.3 ELECTROCHEMICAL INVESTIGATIONS

Corrosion processes are, as a rule, electrochemical in nature. Basic knowledge about the corrosion and corrosion resistance of metals can therefore be obtained from

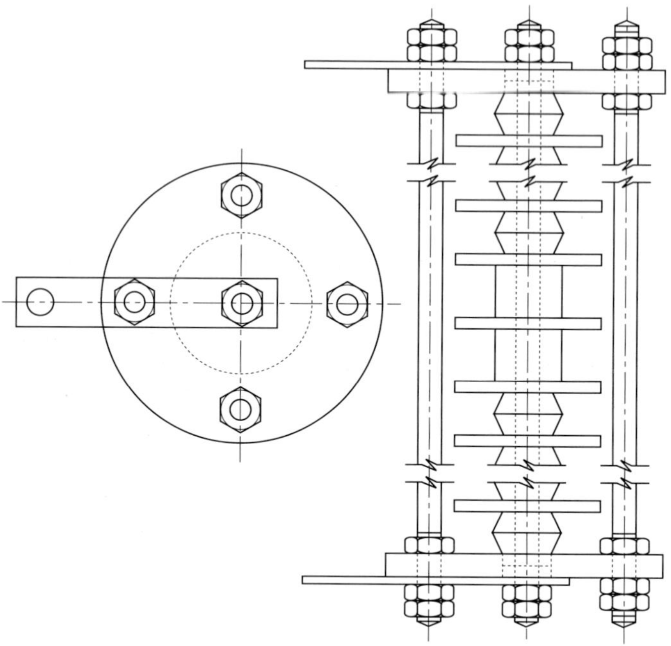

Fig. 149 Rig with test coupons for corrosion testing in tanks, vessels etc. (according to ASTM G4-68).

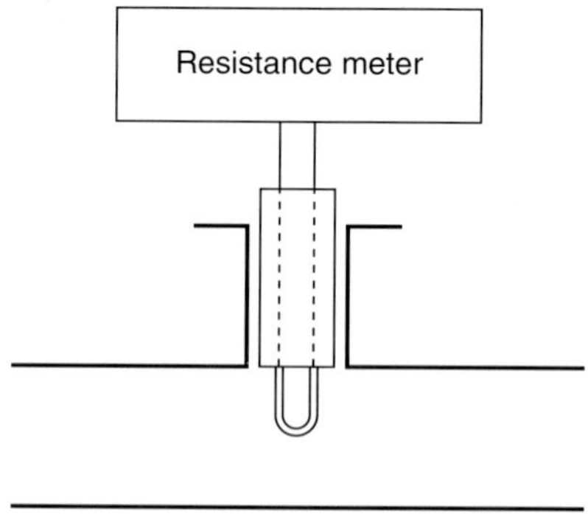

Fig. 150 Resistance measuring probe mounted in a pipe; the probe is often surrounded by a perforated cover to protect from mechanical damage.

electrochemical investigations. These can include determination of corrosion potential and corrosion current as well as anodic and cathodic polarisation curves (see 2.7). Several methods are used for this purpose.

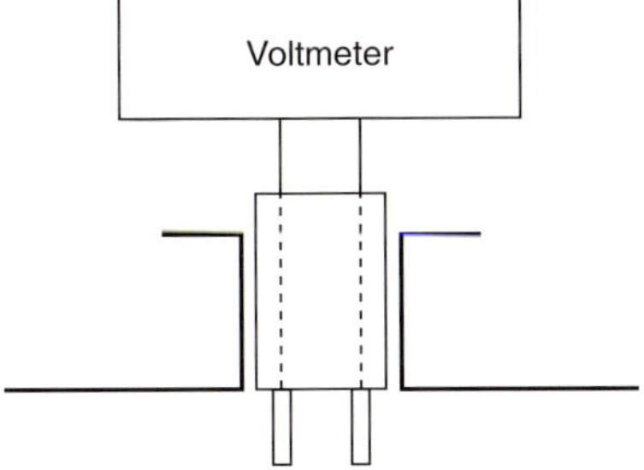

Fig. 151 Probe for the measurement of polarization resistance mounted in a pipe.

Potentiostatic investigations. The electrode potential of the test piece is kept constant, or regulated according to a certain programme while the resulting corrosion current is determined. The current is a measure of the corrosion rate, which however, can also be determined, e.g. from the mass loss. As the corrosion rate is in most cases strongly dependent on the electrode potential of the specimen it is often convenient to carry out exposure testing in aqueous solution under potentiostatic control.

Galvanostatic investigations. The specimen surface is subjected to a controlled current density while the resulting electrode potential is recorded.

Potentiodynamic investigations. The electrode potential of the specimen is changed continuously or stepwise while the corresponding current is recorded. With material which is sensitive to pitting the current will increase significantly when the pitting potential is exceeded (see 8.2).

Polarisation resistance investigations. The so-called polarisation resistance, $R = \eta/i$, is measured in the linear region of the polarisation curve, i.e. close to the corrosion potential (see 2.7). The polarisation resistance is a measure of the inhibition of the corrosion process and is inversely proportional to the corrosion current. Commerical instruments are available for the measurement of polarisation resistance. Here measurement takes place using two or three electrodes assembled together to form a test probe. The result can be read on a scale as corrosion rate.

Electrochemical impedance spectroscopy (EIS) (see 2.7). EIS is used, amongst other things, for studies of corrosion inhibitors and their effectiveness. The curves that are obtained provide information about the nature of electrode reactions although they are not always easy to interpret.

9.4 PHYSICAL METHODS

Corrosion usually leads to changes in the surface condition of a material, e.g. corrosion attack, or the production of a coating of corrosion products or a passivating layer. A number of physical methods are available for the investigations of these changes, e.g. metallographic microscopy, electron microscopy, X-ray or electron diffraction as well as surface physical methods. In Appendix 3 a selection of such methods and their possibilities and limitations is described.

10
Corrosion information

Knowledge about corrosion is very extensive. According to estimates 15% of all corrosion costs could be avoided by the application of what is already known. However, the retrieval of relevant corrosion facts can present problems although various kinds of aid are available.

10.1 CORROSION DATA COLLECTIONS, HANDBOOKS AND STANDARDS

A great deal of knowledge about corrosion is summarised in corrosion data collections, handbooks and standards. A selection of such books is given in Appendix 4.

In corrosion standards one can find, for example, terminology, methods of testing, and quality requirements from the point of view of corrosion and corrosion protection. There are references to such standards in previous chapters. Corrosion standards are published by the International Organisation for Standardisation (ISO), by European Committee for Standardisation (CEN), and by the national standards associations; SIS in Sweden, DIN in Germany, BSI in Great Britain, AFNOR in France etc. Special committees within ISO (TC 156) and CEN (TC 262) have the overall responsibility for corrosion questions. Important corrosion standards are also published by organisations other than the national standards associations, e.g. by American Society for Testing and Materials (ASTM) and by NACE International in the USA.

10.2 JOURNAL ARTICLES AND PATENTS

Research and development work is often published in scientific and technical journals; in Appendix 5 a number of corrosion magazines are listed. In a multidisciplinary area, such as corrosion, much work is published outside the special corrosion journals. To find publications on a specific topic, abstract periodicals or the computer-searchable literature data bases may be consulted. A computer search can be carried out on-line from a local terminal in the data base libraries that are maintained by several organisations, "hosts", e.g. ESA-IRS, STN and DIALOG. A personal computer with external communication capacity is frequently used as the local terminal, and the connection to the host computer is set up via telephone and data transmission networks. The Chemical Abstracts (CA), METADEX and COMPENDEX data bases are frequently useful sources for corrosion literature, although corrosion is only a minor part of their subject coverage. When relevant references have been chosen, the corresponding publications can be ordered on-line, or obtained from public libraries.

Technical inventions are often first reported in patent literature, i.e. in patent applications and in approved patents. Here not only the current invention is described, but frequently useful up-to-date knowledge in the field is also provided as a background.

10.3 CONFERENCES

Conferences provide an opportunity of making informal personal contacts. Some regularly recurring international corrosion events are cited in Appendix 6. Some cover the whole area of corrosion, while others have a more narrow scope. Large national or regional corrosion conferences are arranged annually, in Europe (European Federation of Corrosion), Great Britain (Institution of Corrosion), North America (NACE International) and Australasia (Australasian Corrosion Association).

10.4 DATA BANKS, EXPERT SYSTEMS AND COMPUTERISED CORROSION LIBRARIES

Besides the literature data bases, there are also data banks from which one can obtain various corrosion facts, e.g. potential-pH diagrams or corrosion rates for different metals in specific environments. In addition there are "expert systems", which aim to cover the accumulated human expert knowledge in a given technological area. A characteristic feature of an expert system is often that the user is gradually led to define the problem through decisive questions, which in turn point the system to display the available information on the current question.

Development work on corrosion data banks has been carried out at several institutions, e.g. at DECHEMA in Germany, at NPL (National Physical Laboratories) in Great Britain and at NACE International/NBS (National Bureau of Standards) in the USA. Computer-assisted materials selection and expert systems for corrosion are reviewed by Chawla and Gupta [20].

The great capacity of CD-ROM disks has provided new ways of information presentation, storage and retrieval. A "corrosion library" containing terminology, corrosion data tables, some text books and case histories including diagrams and illustrations, with search facilities, guidance and links between the sections has been issued as a CD-ROM for personal computers [21].

References

[1] *Economic Effects of Metallic Corrosion in the United States.* National Bureau of Standards, *Special Publication 511,* 1978.

[2] Pourbaix, M., *Atlas of Electrochemical Equilibria in Aqueous Solutions.* NACE, Houston, CEBELCOR, Brussels, 1974.

[3] Kucera, V. & Mattsson, E., *Atmospheric Corrosion in Corrosion Mechanisms* (Ed. F. Mansfeld), Dekker, New York, 1987.

[4] Mattsson, E., *Electrochimica Acta*, **3**, 1961, p.279.

[5] Benjamin, L. A., Hardie, D. & Parkins, R. N., *Investigation of the Stress Corrosion Cracking of Pure Copper.* Svensk Kärnbränsleförsörjning AB, Stockholm, KBS Technical Report 83-06,1983.

[6] Kucera, V. & Mattsson, E., *Atmospheric Corrosion of Bimetallic Structures in Atmospheric Corrosion* (Ed. W. H. Ailor), Wiley, New York, 1982.

[7] *Acidification Today and Tomorrow*. Swedish Ministry of Agriculture, Environment '82 Committee, 1982.

[8] Barton, K., Bartonova, Z. & Beranek, E., *Werkst. Korros.*, **25**, 1974, p.659.

[9] Kucera, V., *Ambio*, **5**, 1976, p.243.

[10] Johnsson, T., *Korrosionshärdigheten hos zinkbeläggningar framställda i moderna elektrolytiska bad.* Korrosionsinstitutet, Stockholm, *KI Rapport* 1981:3.

[11] Ericsson, R. & Sydberger, T., *Werkst. Korros.* **31**, 1980, p.455.

[12] Sandberg, B., Private communication.

[13] Thomas, R., *Vamförzinkning*. Nordisk Förzinkningsförening, Stockholm, 1981.

[14] Berg, C. J., Jarosz, W. R. & Salathe, G. F., J. *Paint Techn.*, Nr **510**, 1967, p.436.

[15] Burgmann, G. & Grimme, D., *Stahl Eisen*, **100**, 1980, p.641.

[16] Knotkova, D. *et al.*, in *Atmospheric Corrosion of Metals*, ASTM STP 767, Philadelphia, 1982, p. 7.

[17] *Korrosionstabeller för rostfria stål.* Jernkontoret, Stockholm, 1979.

[18] *Aluminium. Konstruktions- och materiallära.* Metallnormcentralen, Stockholm, MNC handbook nr 12,1983, p. 85.

[19] Scheffer, F. & Schachtschabel, P., *Lehrbuch der Bodenkunde*. Verdinand Enke Verlag, Stuttgart, 1984.

[20] Chawla, S.L. & Gupta, R.K., *Materials Selection for Corrosion Control. ASM International*, Materials Park, 1994.

[21] Agema, K. & Bogaerts, W., *Active Library on Corrosion (ALC).* Elsevier & NACE, Amsterdam & Houston, 1992.

Appendices

Appendix 1. Risk of bimetallic corrosion (BC) in different types of atmosphere [6].

Metal uncoupled: corr rate (μm/year)			Risk of BC	Bimetallic corrosion of metal (in left-hand column) coupled to metals indicated below		
Rur.	Urb	Mar.		Rural atmosphere	Urban atmosphere	Marine atmosphere
Carbon steel						
22	53	36	0:	Pb, Zn, Al, Mg, weath steel	Pb, Zn, Al, Mg, weath steel,	Pb, Zn, Al, anod Al, Cr, Mg, weath steel
			I:	SS, Cu, Ni, anod Al, Sn, Cr	SS, Cu, Ni, anod Al, Sn, Cr	SS, Cu, Ni, Sn
Stainless steel						
0.03	0.04	0.04	0:	Cu, Ni, Cr	Cu, Ni, Cr	Cu, Ni, Cr
Weathering steel						
22	48	32	0:		SS	
			I:	SS		SS
Copper						
1.2	1.4	3.2	0:	C-steel, Pb, anod Al Sn, Cr, Mg	C-steel, Pb, anod Al Sn, Cr, Mg	C-steel, SS, Pb, Ni, anod Al, Sn, Cr, Mg
			I:	SS, Ni	SS, Ni	
Lead*						
0.3	0	0	O:	Zn, Al, anod Al	Zn, Al, Cr	Zn, Al, anod Al
			I:	Ni, Sn, Cr	SS, anod Al, Sn	SS, Cu, Ni, Sn
			II:	C-steel, SS, Cu	C-steel, Cu, Ni	C-steel, Cr
Zinc						
0.5	2.5	1.3	0:	Al, Mg	Al, Mg	Al, anod Al, Mg
			I:	SS, anod Al, Sn, Cr	SS, anod Al, Sn, Cr	SS, Cr
			II:	C-steel, Cu, Pb, Ni	C-steel, Cu, Pb, Ni	C-steel, Cu, Pb, Ni, Sn
Nickel						
0.4	2.7	1.0	0:	Cu	Cu	Cu
			I:	SS, Cr	SS, Cr	SS, Cr
Aluminium						
			0:	Pb, Zn, anod Al, Sn, Cr, Mg	Zn, anod Al, Mg	Zn, anod Al, Mg
0.1	0.5	0.6	I:	C-steel, SS, Cu, Ni, weath steel	C-steel, SS, Pb, Ni, Sn, Cr, weath steel	SS, Ni, Sn, Cr
			II:		Cu	C-steel, Cu, Pb weath steel
Anodised aluminium						
0	0.2	0.2	0:	Pb, Ni, Sn, Mg	Pb, Cr, Mg	Cr, Mg
			I:	C-steel, SS, Cu, Cr	SS, Ni, Sn	SS, Pb, Ni, Sn
			II:		C-steel, Cu	C-steel, Cu
Tin						
1.4	2.7	10	0:	C-steel, SS, Zn, Ni, Al, Cr, Mg	C-steel, SS, Zn, Al, Ni, Cr, Mg	C-steel, SS, Zn, Al Cr, Mg
			I:	Cu	Cu	Cu, Ni
Magnesium						
6.5	11	14	0:	Al, anod Al	Al, anod Al,	Zn, Al, anod Al
			I:	Zn, Sn, Cr	Zn, Sn, Cr	Sn, Cr
			II:	C-steel, SS, Cu Pb, Ni	C-steel, SS, Cu Pb, Ni	C-steel, SS, Cu Pb, Ni

*The results for lead are based on results from a three months' exposure and should, therefore, be considered uncertain.
Key: O: no risk of bimetallic corrosion. I: increased corrosion in couple. II: strongly increased corrosion in couple. Underlining indicates a corrosion rate above 5 μm/year.
anod = anodised.
weath = weathering.

Appendix 2 Hoover's alignment chart for determination of the saturation pH (pH_s) according to the Langelier equation.

Procedure

1. Proceed vertically from the salt content to the prevailing temperature.
2. Proceed horizontally to Line 1.
3. Draw a straight line from the value in Line 1 to the prevailing Ca content in Line 3.
4. Note the value at the point of intersection with line 2.
5. Draw a straight line from this value on line 2 to the prevailing HCO_3^- content Line 5.
6. The point of intersection with line 4 indicates the value of the saturation pH, i.e. pH_s.

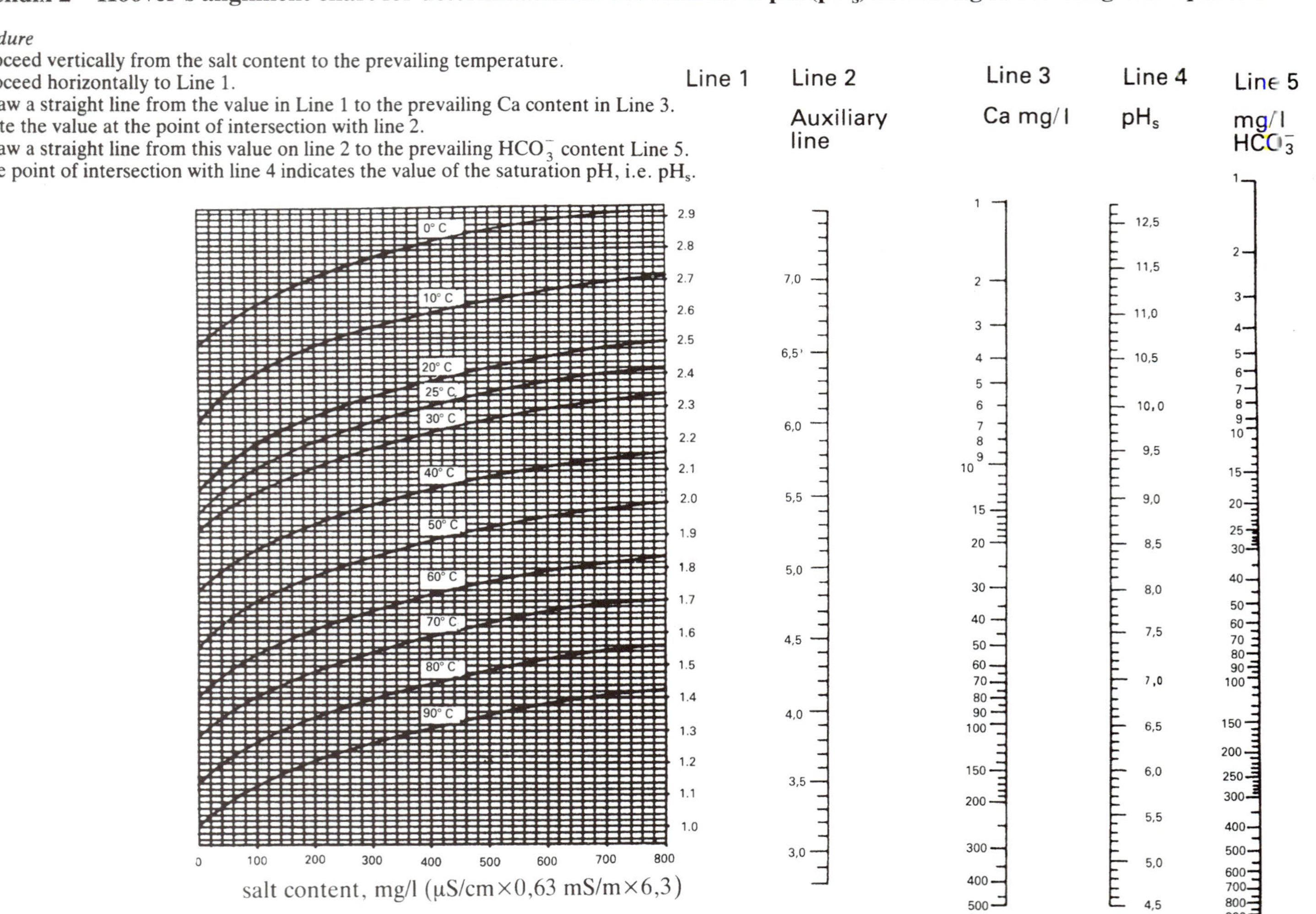

Appendix 3 Some physical methods used in corrosion investigation

Type	Principle employed	Remarks
Optical microscopy	The surface of the sample is examined under a microscope illuminated with visible light.	Magnification: 10–1000×
Scanning electron-microscopy (SEM)	A view of the sample surface is obtained with the aid of a reflected electron beam. X-rays emitted make chemical analysis possible by so-called energy-dispersive X-ray spectroscopy (EDS). The method requires a vacuum.	Magnification: 10^2–10^4×.
Transmission electron microscopy	An electron beam is allowed to pass through a thin sample ($\leqslant 1\ \mu m$) and gives an image of the sample on a fluorescent screen. The diffraction pattern of the radiation makes phase identification possible. The method requires a vacuum.	
Ellipsometry	The sample surface is irradiated with polarised light which is reflected. Changes in the direction of polarisation is a measure of the thickness of the surface coating. The method can be applied to samples in liquids.	Can be used for coating thicknesses >0.1 nm.
X-ray diffraction	A monochromatic X-ray beam is passed through the sample. The diffraction pattern makes phase identification possible.	Can be used on samples of thickness 10–100 μm.
X-ray fluorescence	The sample surface is irradiated with a high-energy X-ray beam. X-ray radiation emitted makes chemical analysis of the surface zone possible.	Detection limits: 10–100 p.p.m. for elements with atomic numbers above 4.
Microprobe	The sample surface is irradiated with a high-energy electron beam. X-ray radiation emitted makes chemical analysis of the surface possible.	Detection limits: 10–150 p.p.m. for elements with atomic numbers above 10.
ESCA or X-ray induced photo-electron spectroscopy (XPS)	The sample surface is irradiated with X-rays. Emission of photo-electrons makes possible chemical analysis of elements in the surface zone and their valency state. The method requires a vacuum.	The analysis is representative for a surface zone of 1–3 nm. Detection limits: ca. 0.1% for elements with atomic numbers above 1.
Auger electron spectroscopy (AES)	The sample surface is irradiated with electrons. Emission of so-called Auger electrons makes chemical analysis of the surface zone possible. Via ion etching the variations in depth can be determined. The method requires a vacuum.	The analysis is representative for surface zones of 0.5–3 nm. Detection limits: *ca* 0.1% for elements with atomic numbers above 2.
Surface-enhanced Raman spectroscopy (SERS)	The sample surface is irradiated with a beam of monochromatic light. The reflected light contains a Raman spectrum which makes possible the identification of ions and bonds in thin coatings , especially on Au, Ag and Cu . The method can be used for samples in aqueous media.	Can be used in the study of thin coatings (<10 nm).
Ion microprobe analysis (SIMS)	The sample surface is irradiated with a beam of ions. Freed molecules are analysed in a mass spectrometer. A depth profile is automatically obtained due to the sputtering effect of the ion beam.	Detection limit in the ppb/ppm region. Applicable for all elements.

Appendix 4: A SELECTION OF HANDBOOKS ON CORROSION

Some of the publications are out of print, but should be available at major libraries.

General

Bockris, J.O'M. & Reddy, A.K.N., *Modern Electrochemistry*. 2 vol., Plenum Press, New York, 1973, 1423 p.

Evans, U.R., *The Corrosion and Oxidation of Metals*. Arnold, London, 1960; First Supplementary Volume. Arnold, London, 1968; Second Supplementary Volume. Arnold, London, 1976.

LaQue, F.L. & Copson, H.R., *Corrosion Resistance of Metals and Alloys*. 2nd ed., Reinhold, New York, 1963, 712 p.

Shreir, L.L. (ed.), Jarman, R.A. (ed.) & Burstein, G.T. (ed.), *Corrosion*. 3rd rev. ed., Vol. 1: *Metal/Environment Reactions*. Vol. 2: *Corrosion Control*. Butterworth-Heinemann, London, 1994, ca 2800 p.

Tomashov, N.D., *Theory of Corrosion and Protection of Metals*. MacMillan, New York, 1966, 672 p.

Uhlig, H.H., *Corrosion Handbook*. Wiley, New York, 1948, 1192 p.

Wranglén, G. *An Introduction to Corrosion and Protection of Metals*. Chapman & Hall, London, 1985.

Corrosion of Metals and Alloys. Terms and Definitions. ISO standard 8044.

Corrosion in different environments

Ailor, W.H. (ed.), *Atmospheric Corrosion*. Wiley, New York, 1982, 1028 p.

Barton, K., *Protection against Atmospheric Corrosion*. Wiley, New York, 1976, 193 p.

Butler, G. & Ison, H.C.K., *Corrosion and its Prevention in Waters*. Leonard Hill, London, 1966, 281 p.

LaQue, F.L., *Marine Corrosion*. Wiley, New York, 1975.

Romanoff, M., *Underground Corrosion*. NBS Circ. 579, US Gov. Print. Office, Washington DC, 1957, 227 p.

Rozenfeld, I.L., *Atmospheric Corrosion of Metals*. English translation by NACE, 1972.

Watkins Borenstein, S., *Microbiologically Influenced Corrosion Handbook*. Woodhead Publishing, Cambridge, 1994, 288 p.

Microbiological Degradation of Materials – and Methods of Protection. Institute of Materials, London, Book B516 : EFC 9, 1992, 84 p.

High temperature corrosion

Kofstad, P., *High Temperature Corrosion*. Elsevier, London, 1988, 546 p.

Kubaschewski, O. & Hopkins, B.E., *Oxidation of Metals and Alloys*. Butterworths, London, 1967.

Lai, G.Y., *High-Temperature Corrosion of Engineering Alloys*. ASM International, Materials Park, 1990, 231 p.

Corrosion protection

von Baeckmann, W.G., Schwenk, W. & Prinz, W., *Handbuch des kathodischen Korrosionsschutzes*. 3rd ed. (in German), VCH, Weinheim, 1989, 545 p.

Bayliss, D.A. & Chandler, K.A., *Steelwork Corrosion Control*. Elsevier, London, 1991, 459 p.

Burns, R.M. & Bradley, W.W., *Protective Coatings for Metals*. Reinhold, New York, 1967, 735 p.

Carter, V.E., *Metallic Coatings for Corrosion Control*. Newnes-Butterworths, London, 1977.

Keane, J.D., *Steel Structures Painting Manual*. Vol. 1: *Good Painting Practice*. 3rd ed. 1994, 649p. - Vol. 2: *Systems and Specifications*. 6th ed. 1991, 459 p., Steel Structures Painting Council, Pittsburgh.

Morgan, J.H., *Cathodic Protection*. 2nd ed., NACE, Houston, 1987, 519 p.

Munger, C.G., *Corrosion Prevention by Protective Coatings*. NACE, Houston, 1984, 530 p.

Nathan, C.C. (ed.), *Corrosion Inhibitors*. NACE, Houston, 1976, 280 p.

Van Oeteren, K.A., *Korrosionsschutz durch Bestrichungsstoffe*. (in German), Carl Hanzer Verlag, Muenchen/Wien, 1980, 1715 p.

Riggs, O.L. & Jr Locke, C.E., *Anodic Protection - Theory and Practice in the Prevention of Corrosion*. Plenum Press, New York, 1981, 284 p.

Rozenfeld, I.L., *Corrosion Inhibitors*. McGraw-Hill, New York, 1977, 327 p.

Wernick, S., Pinner, R. & Sheasby, P.G., *The Surface Treatment and Finishing of Aluminium and its Alloys*. Vol. 1 & 2, 5th ed., ASM & Finishing Publications Ltd, Teddington, 1987, ca 1300 p.

Surface Cleaning, Finishing and Coating - Metals Handbook. 9th ed., Vol.5, ASM, Metals Park, 1982, 715 p.

Corrosion Inhibitors. Institute of Materials, London, Book B559 : EFC 11,1994, 224 p.

Corrosion properties of different materials

Bergman, G., *Corrosion of Plastics and Rubber in Process Equipment - Experiences from the Pulp and Paper Industry*. TAPPI Press, Atlanta, 1995, 181 p.

Britton, S.C., *Tin versus Corrosion*. Tin Research Institute, Greenford, 1977, 180 p.

Davis, J.R. (ed.), *Stainless Steel. ASM Speciality Handbook*. ASM International, Materials Park, 1994, 600 p.

Dolezel, B., *Die Beständigkeit von Kunststoffen und Gummi*. (in German), Carl Hanzer Verlag, Muenchen/Wien, 1978, 684 p.

Godard, H.P , Jepson, W.B., Bothwell, M.R. & Kane, R.L., *The Corrosion of Light Metals*. Wiley, New York, 1967, 360 p.

Gullman, J., Knotkova, D., Kucera, V., Swartling, P. & Vlckova, V., *Weathering Steels in Building - Cases of Corrosion Damage and their Prevention*. Bulletin no. 94, Swedish Corrosion Institute, Stockholm, 1985, 63 p.

Leidheiser, Jr. H., *The Corrosion of Copper, Tin and their Alloys*. Wiley, New York, 1971, 411 p.

Porter, F.C., *Corrosion Resistance of Zinc and its Alloys*. Dekker, London, 1994, 528 p.

Sedriks, A.J., *Corrosion of Stainless Steels*. Wiley-Interscience, New York, 1979, 282 p.

Slunder, C.J. & Boyd, W.K., *Zinc: Its Corrosion Resistance*. 2nd ed. ILZRO, New York, 1983, 250 p.

Corrosion testing

Baboyan, R. (ed.), *Corrosion Tests and Standards. Application and Interpretation*. ASTM, Philadelphia, 1995, 764 p.

Box, G.E.P., Hunter, W.G. & Hunter, J.S., *Statistics for Experimenters*. Wiley, New York, 1978, 653 p.

Champion, F.A. *Corrosion Testing Procedures*. Chapman & Hall, London, 1964, 369 p.

Heitz, E. (ed.), Rowlands, J.C. (ed.) & Mansfeld, F. (ed.), *Electrochemical Corrosion Testing*. VCH & DECHHEMA, Weinheim & Frankfurt/Main, 1986, 352 p.

Corrosion Testing made Easy. Vol. 1: Sedriks, A.J., *Stress Corrosion Cracking Test Methods*. 1990, 87 p. - Vol. 2: Hack, H.P., *Galvanic Corrosion Test Methods*. 1993, 68 p. - Vol. 3: Verink, E.D., *The Basics*. 1994, 200 p. - Vol. 4: Lawson, H.H., *Atmospheric Corrosion Test Methods*. 1995, 77 p., NACE, Houston.

Guidelines for Methods of Testing and Research in high Temperature Corrosion. Institute of Materials, London, Book B604 : EFC 14, 1995, 240 p.

Corrosion data

DECHEMA Werkstoff-Tabelle. (in German), DECHEMA, Frankfurt/Main, from 1953.

Behrens, D. (ed.), Kreysa, G. (ed.) & Eckermann, R. (ed.), *DECHEMA Corrosion Handbook*. Vol. 1-12, DECHEMA & VCH, Frankfurt Weinheim, 1987-1993 (based on DECHEMA Werkstoff-Tabelle above).

Craig, D. & Anderson, B. (eds), *Handbook of Corrosion Data*. 2nd ed., ASM International, Materials Park, 1995, 998 p.

During, E.D.D., *Corrosion Atlas. A Collection of Illustrated Case Histories*. 2nd ed. Vol. 1: *Carbon Steels*; Vol. 2: *Stainless Steels and Non-Ferrous Materials*. Elsevier, Amsterdam, 1988, ca 400 p.

McIntyre, P. (ed.) & Mercer, A.D. (ed.), *Corrosion Standards: European and International Developments*. Institute of Metals, 1991, 112 p.

Corrosion Standards II. National, European and International Standards 1990-1995. Institute of Materials, London, Book B660, 1996, 180 p.

Pourbaix, M., *Atlas of Electrochemical Equilibria in Aqueous Solutions*. Pergamon Press, Oxford, 1966, NACE, Houston, 1972, 644 p.

Schweitzer, P.A. (ed.), *Corrosion Resistance Tables*. 4th ed., 3 vol. (parts A, B, C), Marcel Dekker, New York, 1995, ca 3500 p.

Corrosion Data Survey, Metals Section. 6th ed., 1985, *Non-Metals Section.* 5th ed., 1975, NACE, Houston; also as *CORSOR* data base for personal computers.

Corrosion Handbook: Stainless Steels. Sandvik Steel & Avesta Sheffield, Sandviken, 1994, 112 p.

Stability Constants of Metal-Ion Complexes. Part A. Inorganic Ligands, Part B: Organic Ligands. IUPAC Chemical Data Series, Pergamon Press, Oxford, 1982 and 1979.

Appendix 5. A selection of journals and abstract periodicals for corrosion information

Journal	Publisher	Language
Zairyo to Kenkyo (from 1987 also in English translation *Corrosion Engineering*)	Japan Society of Corrosion Engineers, Tokyo	Japanese
British Corrosion Journal	The Institute of Materials, London	English
Corrosion	NACE International, Houston, Texas	English
Corrosion Australasia	Australasian Corrosion Association, Australia	English

Corrosion Reviews	Freund Publishing House, London	English
Corrosion Science	Pergamon Press, Oxford	English
Farbe und Lack	Kurt Vincenz Verlag, Hanover	German
Journal of Electrochemical Society	The Electrochemical Society, Princeton	English
Materials Performance	NACE International, Houston, Texas	English
Protection of Metals (translation of *Zashchita Metallov*)	Scientific Inform. Consultants, London and Consultants Bureau, New York	English
Materials and corrosion/ Werkstoffe und Korrosion (contains also abstracts)	Verlag Chemie, Weinheim	German
Zashchita Metallov	Akademia NAUK, Moscow	Russian
Corrosion Abstracts	NACE International, Houston, Texas	English
Referatnyi Zhurnal 66; Korrozii i Zashchita ot Korrozii (also in English translation *Corrosion Control Abstracts*)	Akademia NAUK, Moscow (English translation published by Scientific Consultants Ltd and Freund Publ. House, London)	Russian
Chemical Abstracts (also as data base)	American Chemical Society, Columbus, Ohio	English
Metals Abstracts (also as data base)	The Institute of Materials, London & American Society for Metals, Metals Park	English

Appendix 6. Some regular international corrosion events

Conference	Sponsor	Intervals, years
Scandinavian Corrosion Congresses (NKM)	Nordiska Samarbetsgruppen for Korrosion	~3
International Corrosion Congresses (ICC)	International Corrosion Council (ICC)	~3
European Corrosion Congresses Eurocorr	European Federation of Corrosion (EFC)	1
European Symposia on Corrosion Inhibitors	University degli Studi di Ferrara, Italy	5
International Symposia on Corrosion in the Pulp and Paper Industry	NACE International, TAPPI and others	3

Index